New Technologies and the Future of Food and Nutrition

New Technologies and the Future of Food and Nutrition

Proceedings of the First Ceres Conference
Williamsburg, VA, October 1989

GERALD E. GAULL
Vice President, Nutritional Science
The NutraSweet Company

RAY A. GOLDBERG
Moffett Professor of Agriculture and Business
Graduate School of Business Administration
Harvard University

A WILEY INTERSCIENCE PUBLICATION
JOHN WILEY & SONS, INC.
NEW YORK / CHICHESTER / BRISBANE / TORONTO / SINGAPORE

In recognition of the the importance of preserving what has been
written, it is a policy of John Wiley & Sons, Inc., to have books
of enduring value published in the United States printed on
acid-free paper, and we exert our best efforts to that end.

Copyright © 1991 by John Wiley & Sons, Inc.

Library of Congress Cataloging in Publication Data:

Ceres Conference (1st : 1989 : Williamsburg, Va.)
    New technologies and the future of food and nutrition :
  proceedings of the First Ceres conference, Williamsburg, VA, October
  1989 / edited by Gerald E. Gaull, Ray A. Goldberg.
        p.   cm.
    Includes bibliographical references.
    ISBN 0-471-55408-1 (alk. paper)
    1. Food—Biotechnology—Economic aspects—Congresses.
    2. Nutrition policy—Congresses. 3. Agricultural biotechnology—
    Economic aspects—Congresses. I. Gaull, Gerald E. II. Goldberg,
    Ray Allan, 1926-   . III. Title
    TP248.65.F66C47   1989
    664—dc20   91-18915
                                                            CIP

Printed in the United States of America

10 9 8 7 6 5 4 3 2 1

Contributors

Philip W. Anderson, Ph.D.
Princeton University
Department of Physics
Princeton, NJ

Kenneth J. Arrow, Ph.D.
Department of Economics
4th Floor, Encina Hall
Stanford University
Palo Alto, CA

Durward F. Bateman, Ph.D.
College of Agriculture
North Carolina State University
112 Patterson Hall
Raleigh, NC

Eneko de Belausteguigoitia, Ph.D.
Alfa Elai, S.A.
Bosque de Ciruelos #304-7
Mexico City, Mexico

Daniel Bell, Ph.D.
American Academy of Arts and
Sciences
Norton's Woods
136 Irving Street
Cambridge, MA

D. Theodore Berghorst, MBA
Vector Securities International, Inc.
1751 Lake Cook Road, Suite 350
Deerfield, IL

Winston J. Brill, Ph.D.
Winston J. Brill & Associates
4134 Cherokee Drive
Madison, WI

George E. Brown, Jr.
Committee of Science, Space &
Technology
2300 Rayburn House Office Building
Washington, DC

Sir Roy Denman
Denman & Partners, LP
2550 M Street, NW, Suite 460
Washington, DC

John T. Dunlop, Ph.D.
Harvard University
Littauer 208
Economics Department
Cambridge, MA

Gerald E. Gaull, M.D.
The NutraSweet Company
1751 Lake Cook Road
Deerfield, IL

Francis Gautier
BSN Groupe
7, rue de Teheran
75381 Paris Cedex 08, France

Donald A. Glaser, Ph.D.
University of California at Berkeley
Berkeley, CA

Ray A. Goldberg, Ph.D.
Harvard Business School
Baker Library 236
Boston, MA

Jules Hirsch, M.D.
The Rockefeller University
Rockefeller Hospital
66th & York Avenue, 2nd Floor
New York, NY

Arthur Kornberg, M.D.
Stanford University Medical Center
Dept. of Biochemistry, Bldg. 415
Stanford, CA

Norman Kretchmer, M.D., Ph.D.
University of California at Berkeley
Dept. of Nutritional Sciences
119 Morgan Hall
Berkeley, CA

Philip Leder, M.D.
Harvard Medical School
Dept. of Genetics
25 Shattuck Street
Boston, MA

Michael A. Miles
Kraft General Foods
Kraft Court
Glenview, IL

Sanford A. Miller, Ph.D.
Graduate School of Biomedical Sciences
The University of Texas Health
*   Sciences Center at San Antonio*
7703 Floyd Curl Drive
San Antonio, TX

John A. Moore, DVM
Institute for Evaluating Health
*   Risks*
100 Academy Drive
Irvine, CA

Irwin H. Rosenberg, M.D.
Center on Aging
Tufts University
711 Washington Street
Boston, MA

Judith S. Stern, Ph.D.
University of California at Davis
Department of Nutrition
Food Intake Laboratory
Davis, CA

Louis W. Sullivan, M.D.
Department of Health & Human
*   Services*
200 Independence Ave. SW, Room
*   615F*
Washington, DC

M.S. Swaminathan, Ph.D.
M.S. Swaminathan Research
*   Foundation*
14, II Main Road
Kottur Gardens
Kototurpuram, Madras, India

R. Thomas Vyner
J. Sainsbury Ltd.
Stamford House
Stamford Street
London SE1 9LL, England

Clayton Yeutter, J.D., Ph.D.
U.S. Department of Agriculture
Washington, DC

Frank E. Young, M.D., Ph.D.
Food and Drug Administration
Room 1471
5600 Fishers Lane
Rockville, MD

Contents

Acknowledgments xi

Foreword xiii
John T. Dunlop

Introduction: New Issues Confronting the Food Industry 1
Gerald E. Gaull

PART I. DEMOGRAPHICS, AGRICULTURE, AND THE
FOOD INDUSTRY 11
Abstract

Chapter 1. Demographic Trends in Populations, Food,
and the Environment 13
Daniel Bell

Chapter 2. Paradigms of American Agriculture 19
Durward Bateman

Chapter 3. Recent American Agricultural History 25
George Brown

Chapter 4. Technological Advances and World Hunger 31
M. S. Swaminathan

PART II. THE BIOLOGICAL REVOLUTION, AGRICULTURE,
AND THE FOOD INDUSTRY 37
Abstract

Chapter 5.   Introduction: Biotechnology and Industry   39
              Donald Glaser

Chapter 6.   A Primer on Molecular Biology of DNA: Its
              Modifications and Role in Heredity   41
              Arthur Kornberg

Chapter 7.   Plant Genetic Engineering and the Food
              Industry   45
              Winston Brill

Chapter 8.   Genetically Engineered Animals   49
              Philip Leder

PART III. THE FUTURE OF NUTRITION   55
          Abstract

Chapter 9.   Understanding Life and Nutrition as Chemistry   59
              Arthur Kornberg

Chapter 10.   A Brief History of Nutrition: The Role of
              Clinical Science   71
              Jules Hirsch

Chapter 11.   The Evolution of Diet and Nutrition   75
              Norman Kretchmer

Chapter 12.   Nutrition and the Elderly   79
              Irwin Rosenberg

Chapter 13.   Nutrition and the Consumer   83
              Judith Stern

PART IV. FOOD, NUTRITION, AND HEALTH: THE ROLE
         OF GOVERNMENT   89
         Abstract

Chapter 14.   The Politics of Food and Biotechnophobia   91
              Clayton Yeutter

Chapter 15.   The Link between Diet and Health   97
              Louis W. Sullivan

Chapter 16. Regulatory Challenges and Biotechnological
             Advances   101
             Frank E. Young

Chapter 17. Food Regulation: Future Issues   105
             Sanford Miller

Chapter 18. The Need for a Scientifically Literate Public   109
             John Moore

PART V. THE FORCES AFFECTING FOOD:
         IMPLICATIONS FOR INDUSTRY   111
         Abstract

Chapter 19. European Community Trade Issues Affecting
             the Food Sector in the 1990s   113
             Sir Roy Denman

Chapter 20. The Food Industry and the U.S. Consumer   119
             Michael Miles

Chapter 21. The Food Industry in Britain   121
             R. T. Vyner

Chapter 22. Investment and the New Technologies   123
             D. Theodore Berghorst

Chapter 23. The Future of the Food Industry in Mexico   127
             Eneko de Belausteguigoitia

Chapter 24. The Food Industry in France and the European
             Community   129
             Francis Gautier

PART VI. ECONOMIC PREDICTION   133
         Abstract

Chapter 25. Economic Forecasting   135
             Kenneth Arrow

Chapter 26. Perception and Predictions   141
             Philip Anderson

Conclusions: The Impact of New Technologies and the Future of the Global Agrifood System in the 21st Century   149
Ray A. Goldberg

Biographies of Contributors   163

Index   169

Acknowledgments

W e are grateful to Robert B. Shapiro, former Chairman of the Board and Chief Executive Officer, and to Lauren Williams, President and Chief Operating Officer, The NutraSweet Company, who abetted the project and provided valuable guidance, and to Max Downham, Vice-President, Strategy and Planning, and his staff who were responsible for the execution of so much that was needed to launch and to execute the Ceres Conference. We are also grateful to members of the advisory committee who provided help in planning: Professors Donald Glaser, Norman Kretchmer, Jules Hirsch, and Sanford Miller, and Messers. Robert Craig and William Ruder. The participants, from both the academic and scientific world and from the business world, made the event memorable.

Foreword

This volume, and the Ceres Williamsburg Conference of October 22–24, 1989 that it reports, is a penetrating account of emerging features in the food industry worldwide, the projected consequences of these developments, and a statement of the private and public issues of policy that emerge and cry out to decision-makers.

Ceres was the goddess of agriculture in Roman mythology, and for this conference as in ancient Rome, is used to symbolize the function of ensuring an abundant and ample harvest. This volume readily projects the triumph in our day of Ceres over the scourges of Malthus. "Biotechnology is the most important scientific tool to affect the food economy in the history of mankind...The half trillion dollar agribusiness economy of 1950 [worldwide] will become a $10 trillion economy by 2028." (See Conclusions by Professor Goldberg.)

The editors provide an analysis of key developments that are shaping the agrifood sector and the food chain of the future and that generate issues for private and public sector policies. Among the major factors are the following:

- Biological revolution and genetic engineering.
- Globalization of economies, politics, and communications.
- Nutrition and health—in the developing world, food scarcity, and, in advanced countries, too much food and the wrong kind of food.
- Population pressures in countries with immense populations and shrinking farmlands.
- Food regulation problems arising from new technology and worldwide distribution.
- Intellectual property rights associated with the new technology.
- Public reactions and communications among major participants in the new agrifood industry.

This listing of factors affecting the agrifood sector and the consequent emergent changes required in private business and public organizations and their policies, in turn, raise a variety of other questions:

- Is the new technology size neutral or partial to large-scale producing and distributing units?
- From whence is the risk capital to come for the research and development given the history of agricultural research and extension services?
- Where are the new leaders of the agrifood industry to be trained?
- How is the displacement of people to be cushioned as changes take place brought on by technology and market shifts?
- How is the credibility of the new technology and agrifood industry with the public and political process to be achieved?

This volume develops the facts and tendencies for the future and explicates the complex issues that lie in wait for private and public decision-makers. That the editors and authors have done extraordinarily well. But deliberately they have left to another conference and volume the critical questions of what is to be done about these issues and who is to take the lead. How does the world community manage these gigantic developments? How does it harness the new technology, and handle its side effects, for the benefit of world populations? How are the concerns with public safety to be handled in a new regulatory regime, and how is public confidence to be won? What are the related roles of private business and public agencies, and what new institutions and partnerships are required for the agrifood sector of the 21st century?

The twin forces of fundamental technological change and globalization of markets are combining in our times to alter forever many of the traditional elements of our economy, sometimes in a dramatic eruption or more often at a measured pace. The food chain and the related pharmaceutical sphere, at the threshold to the 21st century, are among the most significant and pivotal sectors to be invested with these forces, but they are not the only ones. We see the same twin factors at work in telecommunications, clothing, transportation, and entertainment, activities no less basic to mankind.

In my experience the management of change in these circumstances, or even the deliberate adjustment to change, requires a continuing forum with all the consequential organizations for periodic discourse, for establishment of the facts and their interpretation, and for the encouragement of research focused on the issues and appropriate feedback. All parties, private organizations and public agencies alike, learn in the process, and new forms of partnership or private–public coalitions tend to emerge.

Indeed, an older, often obsolete governance system comes to be displaced or evolves into a new one. I have seen these consequences emerge from many forums, and I believe the agrifood sector and the management of the changes induced by the biotechnology revolution and globalization require such a continuing medium of discourse, research, problem-solving, and negotiations.

JOHN T. DUNLOP

Introduction: New Issues Confronting the Food Industry

Gerald E. Gaull

In October, 1989, a group of influential thinkers, policy makers, and food industry executives from around the world gathered in Williamsburg, Virginia, for the first Ceres Conference. The purpose of the conference was to identify the key scientific, socioeconomic, and political issues and questions affecting the food industry as it enters the 21st century. In the past, these issues usually have been considered separately: scientists have talked about science; food industry executives about food products; government officials about food policy and regulation. And each sector has come to speak in its own language, further isolating itself from the others.

It was time to bring together in one room a group of serious thinkers from each of these sectors and to see whether or not there was agreement, at least about the major issues and questions affecting the agrifood industry and its ability to thrive in the next century. There was little expectation that precise answers would be found; rather, the purpose was to define the issues and to frame the questions. The result was a new appreciation of the complexity of these issues and a clearer articulation of the questions. The actionable business answers, of course, will come later and will probably be different for each company. This book grew out of that unprecedented meeting, and it attempts to convey to the reader some of the most important ideas and discussion of them that took place.

GLOBALIZATION

As the agrifood business moves into the 21st century, it must cope with a rapidly changing business environment of great complexity. One of the most important trends affecting it is "globalization." Politics, economics, trade, and communications operate at an increasingly global level, making it necessary for food industry executives to have a deeper understanding of foreign cultures than ever before. They must comprehend the political system, business practices, consumer needs and food safety regulations of each country in which they do business.

Forecasting political change is too complex a task for social scientists, but one trend is clear: the nation-state is becoming too small to solve the big problems facing the food industry. Even the United States, with its tremendous capacity for producing food, has seen the dollar value of food imports double over the last two decades. As long as each country has its own complex set of food safety regulations, for example, free trade will be impossible. There is a critical need for international bodies, vested with real power, to regulate food safety, the environment, and trade.

THE BIOLOGICAL REVOLUTION

In the midst of these changes, a revolution in biological science is occurring, with a potential impact on society rivaling that of the changes in physics that took place at the beginning of the 20th century. Since 1953, when James Watson and Francis Crick elucidated the structure of DNA, the molecule which encodes hereditary information, scientists have succeeded in pinpointing more and more individual genes, steadily improving their understanding of genetic mechanisms. Last year, the federal government launched the Human Genome Project, a massive effort to make a detailed genetic map of all the human chromosomes. This initiative is of great importance both to genetic scientists, who are developing new techniques for transplanting genes from one organism to another, and to nutritional scientists, who then will be able to gain invaluable insights about the nutritional requirements of individuals and groups, as opposed to entire populations.

The new set of techniques, known collectively as genetic engineering, opens up dramatic possibilities for agriculture. Biologists can now create plant varieties that will provide bigger harvests and improved nutritional quality. Plants can be made resistant to insects and other crop-destroying organisms so that pesticides will not have to be used, reducing both cost to the farmer and health concerns of the consumer. Corn, which is deficient in the essential amino acid lysine, can be made into a higher quality protein, a boon for people in some developing countries who depend heavily on corn as the main staple of their diet. The list of possibilities is endless—cotton with better fiber qualities, genetically engineered wheat for types of flour better for particular uses, tomatoes that are tough enough to be shipped to supermarkets but don't taste like cardboard.

The impact of genetic engineering on animal breeding is equally impressive. Genetic changes that normally take decades of traditional breeding can now be accomplished in just a few years. Improved milk production is one example. Through classical cattle breeding and improved nutrition, the average milk yield in the United States increased from 4600 pounds per cow in 1941 to 14,000 pounds today. Using techniques of genetic engineering, predictions have been made that 50,000 pounds per cow could be achieved in the next 5–25 years. In a truly free market, this will result in significant price reductions to the consumer. Genetic engineering also will produce animals that are themselves healthier, by introducing genes for disease resistance, and more healthful for man, for example, by introducing genes that increase the ratio of lean meat to fat.

Clearly, biotechnology has tremendous potential for our society, helping farmers, food processors, and consumers. But in the developing world, where food production is not keeping pace with a burgeoning population, biotechnology holds out an even greater promise—the elimination of mass hunger. For this to happen, however, the new technology must be applied in a way that meets the specific problems faced by such countries. Furthermore, solutions providing equity to both the developed and the developing worlds must be found in order to ensure business interest and continuing investment.

THE NEW DEMOGRAPHICS

In Asia, Africa, and Latin America the rate of increase in the population is said to be slackening, but it is still increasing and more than 40% of the population is under 17 years old. In the next generation, the population of the developing countries has been predicted to double. At the same time, however, the amount of land usable for farming is decreasing in these countries. Soil erosion, falling water tables, and diminishing returns from chemical fertilizers and pesticides have all taken their toll on land productivity. Countries such as Mexico, India, and Indonesia, which became self-sufficient in food production in the 1960s, thanks to new farming methods, are now sliding back to a situation in which population growth is outstripping food supply.

Mexico is a case study of the problems that the developing world will face in the 21st century. Its population of 85 million is expected to double in 20 years, since 42% of Mexicans today are under the age of 15. But Mexico's agriculture is in crisis, and millions of farmers are leaving their rural homes and streaming into the cities. By the year 2000, Mexico City is expected to have a population of 30 million. Meanwhile, United Nations' studies indicate that more than half of all Mexicans now live below minimal national standards. In the countryside, four out of five children under 4 years old are below the normal weight and height.

It is precisely in countries like Mexico, India and Indonesia, with their immense populations and shrinking farmland that the biological revolution has its greatest potential for social benefit. Plants can be made genetically resistant to drought and to soil imbalances like salinity and alkalinity. Moreover, biotechnology makes it possible to create value-added products from parts of the plant not used directly for human food, for example, food for stall-fed animals.

NUTRITION AND HEALTH

While the central nutritional problem in the developing world is food scarcity, that in the industrialized world is too much food or the wrong kinds of food. Obesity affects 34 million Americans resulting in a much greater likelihood of developing heart disease, hypertension, and diabetes. Here, the task for the 21st century is to improve our understanding of nutrition in order to produce the right foods from which to select a healthful diet. With all the excitement over advances in genetics, the crucial field of nutritional science has been relatively neglected. It is difficult to attract federal grants and young scientists for work in this important area. The result is that we now know much more about DNA than about the mechanisms controlling utilization of nutrients. And, much of our public policy is based on nutritional epidemiology which can only suggest, but cannot establish, the causal relationships between nutrition and health.

The debate over cholesterol is a case in point. A few years ago, scientific evidence suggesting that a diet low in cholesterol reduces the chances of heart disease brought about a change in the eating habits of millions of Americans, and a myriad of new low-cholesterol food products in the supermarkets. But the cause of heart disease is far more complex, including factors such as saturated versus unsaturated fatty acids and fiber, stress, and the genetic proclivity of individuals. Recently, some have challenged the scientific data on which dietary guidelines for the general population are based. Products made with oat bran, once snapped up by cholesterol-conscious consumers as fast as the food companies could produce them, are spending more time on supermarket shelves.

Bewildered by conflicting news reports and an arsenal of supermarket products making carefully worded health claims, consumers are losing faith in the scientists who make dietary recommendations, as well as in the companies that manufacture the foods. The consumers need to understand that the science of nutrition is evolving and that current dietary recommendations should be understood as reasonable things to be doing in the light of incomplete information. Dietary guidelines, and the health claims emanating from them, are too often taken to mean that if a consumer eats certain foods, he or she will not get cancer or heart disease. These relationships, however, are not hard and fast, and to the extent that they are statistical truths, they are applicable only to specific groups and not to individuals.

Part of the problem is that individual differences in body chemistry make it extremely difficult to generalize about nutrition in a population. A particular food is not good or bad per se. It is now becoming increasingly clear that a particular food must be seen in the context of a complete diet, of a person's lifestyle, and of the genetic makeup of the individual who lives on that diet. Some people can handle more saturated fats and salt in their diet; others can handle less or only one of the two; still others can handle little of either. Millions of Orientals, African–Americans, and Southern Europeans are unable to drink milk beyond the first few years of life, because they lose lactase, the intestinal enzyme needed to digest lactose, the sugar contained in all milks drunk by man. The elderly, a rapidly growing segment of the population in the United States, Europe, and Japan, have their own special dietary requirements. However, current recommended dietary allowances (RDAs) in America do not distinguish between people 55 years old and those older. Clearly, a quintogenarian has different dietary needs from an octogenarian. But, how much more or less and for which nutrients?

The challenge facing nutritional science is to understand these individual differences well enough to create specific nutritional profiles. The food industry will then have the opportunity to respond by making products that are tailored to the dietary needs of various groups of people. For example, lactose-free milk is already on the market. The supermarket of the future may have special sections and products not only for dieters, but for diabetics, lactose-intolerants, and other large groups with special dietary needs.

THE COMMUNICATION GAP

As great as the potential benefits of biotechnology are, they will not be realized unless certain key obstacles are overcome. Perhaps the biggest of these is the widening gap between our burgeoning scientific and technological capabilities and the public's understanding of them. There is widespread public fear of these new advances, especially genetic engineering, and its consequences for the food industry.

Critics have emerged whose political agenda includes a moratorium on biotechnological research and production. They prey upon public anxiety that recombinant DNA research could create strange and uncontrollable life forms or new diseases, that is, "biotechnophobia." In fact, the best a molecular biologist can do is to add a few genes to the hundreds of thousands that already exist in an organism. The idea that one can splice half the genes of one organism together with half the genes of another and expect to produce anything that lives is pure science fiction.

The "green movement" has become an increasingly potent political opponent of genetic engineering. In West Germany, the "greens" have successfully stymied agricultural field testing and production of genetically engineered pharmaceutical products. Frustrated German companies are setting up research facilities in the United States. Human insulin, produced by recombinant DNA bacteria, could not be made in Germany until very recently. Instead, the Germans imported it, even though they long had the capability of making it themselves. The influence of such ideology is felt throughout Europe and to a lesser degree in the United States.

Of particular concern to the food industry is the public's fear of consuming the new products coming from genetic engineering. For example, the genetically engineered growth hormone, bovine somatotropin (BST), has created a major controversy in this country. When administered to cows, BST increased milk production by as much as 25%. Despite the Food and Drug Administration (FDA) opinion that milk from BST-treated cows is safe, certain public advocacy groups raise fears, which are amplified by the news media. Not only is BST attacked as unsafe, but as economically undesirable. Small dairy farmers, it is argued, would be driven out of business by a flood of cheap milk. The benefits of cheap milk to consumers, particularly low-income consumers, is never mentioned. The result is a public backlash, culminating in recent bans on milk produced from BST-treated cows in Wisconsin and Minnesota and the well-advertised refusal of some dairy product processors and supermarket chains to sell products from cows so treated.

It is in the nature of contemporary life, and in particular the media, that new concerns must be raised periodically. Food, being very close to our concerns about health and life itself, has been a major candidate for continual alert. Any scientific report, no matter how preliminary, which suggests a link between a food additive and a disease, becomes cause for widespread alarm. The 1989 scare over Alar in apples is an egregious example of incited public skittishness over a low-risk additive. The risk of innovation is often trumpeted by consumer advocates abetted by the media; this along with a call to return to the by-gone halcyon days of safety. Current discussion focuses, in the words of Wildavsky, "almost entirely on the dangers of risk taking while neglecting, to the detriment of our common safety, opportunity benefits that would be lost by risk aversion."

But while irrational public fears whipped up by the media are clearly part of the problem, it is also true that neither the food industry nor the scientific community is doing enough to educate the public about biotechnology. Scientists can no longer retreat to their laboratories nor food executives to their offices. Everyone with an interest in human progress must somehow join in the effort to educate the public about the benefits of the new technology. Clearly, not enough has been done. But, who will take the lead?

FOOD REGULATION

Food regulatory agencies like the FDA have the difficult job of weighing inconclusive scientific evidence, along with public perceptions and business realities. As the first products of genetic engineering enter the marketplace, the challenges facing regulators are greater than ever.

Rather than biotechnology, pesticides, or food additives, the main real food safety issue of the 21st century is likely to be microbial contamination. Already, 33 million Americans a year suffer various degrees of gastrointestinal illness caused by microorganisms in food, and every year the world's annual mortality due to infectious diarrheal disease alone is at least 4.6 million. Food safety experts are calling for a broader effort to inspect food worldwide and to inform the public about sanitary methods of food preparation and handling. Yet, food irradiation, a powerful and safe antimicrobial tool, has been abandoned in the United States because of irrational public fears (see subsequent discussion).

Meanwhile, the globalization of the food industry creates new regulatory problems. Many countries keep out U.S. food products on the basis of safety concerns and regulations. A European ban on steroid-hormone treated cattle provoked charges of protectionism from American beef producers and trade officials. This problem underscores the need for international regulatory standards for food safety.

The advent of genetic engineering poses a host of new questions with which regulators have only begun to grapple. Do we need to limit the number of genes that can be implanted into a particular type of plant or animal? How many new genes can you add without compromising safety? A single plant breeder may introduce very many new genotypic plant varieties every year. Since regulating plant by plant is out of the question, regulators will have to develop broad categories to determine which products need to be scrutinized.

TECHNOLOGY TRANSFER

Intellectual property rights are another issue raised by the biological revolution. When a new plant or animal species is engineered by a team of genetic scientists, can it be patented? In 1988, Harvard University was granted the first U.S. patent for a genetically engineered animal, a tumor-prone mouse to be used in the study of cancer. But many countries do not recognize patents for living organisms. Will they have to in the future? And if so, who receives the profits—the corporation, the university, or the country that provides the genetic raw materials?

Over the past 50 years the developing world, with its vast rain forests, has been the source of many of the genes used by Western biologists to improve agricultural species and develop new medicines. A large portion of all the world's plant and animal species live in the tropical rain forests. The nations of the developing world want a share of the revenues generated by the germ plasm coming from their indigenous flora and fauna. On the other hand, the "gene-poor" nations of the West stress the large investments in research they must make in order to reap the benefits of biotechnology. These are high-risk business ventures, and one thing is absolutely clear: the research and development costs of these new technologies are large and long term. Investment money will disappear if innovation is not suitably rewarded. While this dispute continues, American biologists on collection missions are becoming increasingly unwelcome in many developing countries. Clearly, some equitable agreement must be reached by which developing countries share in the benefits of biotechnology.

THE NEW AGRIFOOD INDUSTRY

As a new multitrillion dollar agrifood industry emerges in the 21st century, the issues of public acceptance, economic impact on agriculture, intellectual property rights, and trade policy must all be reckoned with. How will the economic benefits of these new technologies be apportioned among the various parties—corporations, farmers, consumers, the industrialized nations and the developing world?

Perhaps it is premature to begin dividing up the profits. History shows us that it takes decades for industry and the consuming public to figure out how to use a new technology to its best advantage. The rapid advances in information technology of the last 15 years is only now being translated into improved productivity in the service sector. For example, the E-mail computer networks first devised 20 years ago are just now coming into general use, and there are now about 40 million personal computers in America, compared with 1 million just 10 years ago. We now can order our groceries at a nearby supermarket via computer!

Ultimately, it is the business community that must decide what new technologies and products are worth developing. In the case of biotechnology, food industry executives must make decisive, long-term investments in the face of global competition, rapid scientific change, and uncertain markets. At the same time they face the pressures of maintaining near-term profit.

CONCLUSION

There is clearly a mutuality of interest among the various public and private sectors in closing the ever-widening gap between burgeoning scientific and technological know-how and the public's ability to understand and accept it. This is one of the central issues facing the agrifood industry, and the various public and private sectors must share responsibility in coping with this problem. As a corollary, the public is confused about nutrition, a relatively neglected science compared with the glamourous attraction afforded the biomedical scientist by the wonders of biotechnology.

Most important, industry must find new ways to assure the public that the new food technologies are not only beneficial to society, but safe. This means making sure that scientists, regulators, public policymakers, and industry leaders continue the discussions that began at the first Ceres Conference. Perhaps it means creating new institutions that can speak to the public with one voice and with unquestioned probity.

BACKGROUND REFERENCES

Bell D. The world and the United States in 2013. *Daedelus* 1987 Summer.

Brown LR. The changing world food prospect: the nineties and beyond. In: *State of the World*, 1989. Washington DC: Worldwatch Institute; 1989; Worldwatch Paper 85.

Ingersol B. Tide of imported food outruns FDA ability to spot contamination. *Wall Street Journal*. 1989 September 27;sect A:1.

Rhoter LK. Mexico feels squeeze of years of austerity. *The New York Times*. 1989 July 25;sect A:1.

Roberts L. Genetic engineers build a better tomato. *Science* 1988;241:1290.

Rosen JD. Much ado about Alar. *Issues in Science and Technology*.1990;Fall:85–90.

Snyder JD, Merson MH. The magnitude of the global problem of acute diarrhoeal disease: a review of active surveillance data. *Bull WHO* 1982;60:605–13.

Stevens WK. Officials call microbes most urgent food threat. *The New York Times*. 1989 March 28;sect C:1.

Tannahill R. *Food in History*. New York: Crown Publishers, Inc. 1988.

American Association for the Advancement of Science. *Science*. 1989;244: 1225–1412.

DEMOGRAPHICS, AGRICULTURE, AND THE FOOD INDUSTRY

ABSTRACT

Harvard sociologist Daniel Bell identifies some demographic trends with important implications for the food industry. Overall world population growth is only part of the problem. The more serious problem is the one of age and food distribution. While the populations of the United States, Japan, and Europe are aging, the reverse is happening in Latin America, Africa, and Asia, where between 40–50% of the populations are under the age of 15. In the next 10 years, a tidal wave of young people from these regions will be flooding the world. Who will feed them, and who will employ them so that they can feed their children? Bell argues that even though world food production is adequate to feed an increased world population, it is not reaching those that are not producing food or have the resourses to purchase it. One-third of the people in the developing world are seriously malnourished. In Bell's view, this suggests that the food problem is not an agricultural or technological one, but rather a problem of politics and socioeconomic organization.

The funding and control of agricultural research will be another key issue in the 21st century. Durward Bateman, Dean of the College of Agriculture at North Carolina State University, is concerned that lack of government support for new agricultural technology will reduce the technological leadership of the United States in the global food system.

Congressman George Brown of California, a member of the House Agriculture Committee and chairman of the Science, Space and Technology Committee, warns that neglecting the needs and aspirations of developing countries is a recipe for political and economic upheaval. The increased efficiencies possible through biotechnology may be viewed as a threat to agricultural workers throughout India, China, and other developing countries. Brown reminds us that the citizens of developing nations constitute a huge pool of potential talent, as well as a future market for the products of the developed countries and that the positive and negative effects of technology have to be taken into consideration in developing technology policy.

M.S. Swaminathan, President of the International Union for the Conservation of Nature and Natural Resources, raises some of the concerns of developing countries as they face the challenge of feeding more and more people from a shrinking base of arable land. What the developing world needs most is agriculture that is land-saving and animal husbandry that is grain-saving. As the limitations of pesticides of the green revolution become more apparent, these countries also need genetically engineered plants that are resistant to the pests and pathogens that are rampant in tropical climates. Moreover, Swaminathan argues that the gene revolution must be implemented in such a way that it not only feeds more people in the developing world, but creates jobs for them as well. If scientists from the developed countries are going to utilize the rain forest of the developing countries for new gene plasm, they must share their findings in an equitable way that improves the welfare of participants in the world food system and the many diverse customers served by that system.

CHAPTER 1

Demographic Trends in Population, Food, and the Environment

Daniel Bell

The basic framework of demography is central to our understanding of society and of agriculture. If one looks at the earlier apprehensions regarding the total growth of population, such concerns probably can be allayed. Many years ago, people were concerned with the notion of "the population bomb" and that somehow the planet would be engulfed completely by the year 2000 or the year 2020.

As in many growth rates, as one might have suspected, the exponential rate has been leveling off. We have passed the point of inflection, so the rates are slowing. The absolute numbers are increasing, but by and large the population growth rates are slowing, and therefore the kinds of earlier apprehensions about being totally swamped by people may not come true.

More important, however, is not the overall level but the disaggregations. We have a bimodal distribution, meaning that in the developed countries of the world, that is, the United States, Europe, and Japan, we have an aging population. We have a situation where, increasingly, a large proportion of the society is becoming older and therefore living longer. In Japan present projections indicate that by the year 2020 or 2030, the ratio of working-age population to old people will be almost one to one; this kind of problem will occur in Germany, western Europe, and the United States as well, perhaps not in that exact ratio, but the effect will be that of a smaller number of working-age people supporting a larger population.

The crucial question, however, is not the aging issue. That problem is somewhat manageable within the framework of the economic development of the advanced societies. The basic factor is that in most of the world today, that is, in Latin America, Africa, and Asia, the proportion of population under age 15 comprises between 40% and 50% of the total population, an accumulation of young people. In North Africa approximately 42% of the population is under 15. In Algeria, the number rises to 46%, and in all of Africa, the proportion under 15 is between 45% and 47%. In the Middle East the proportion is even higher, with approximately 49% of the population under 15. In southern Asia the number is about 40%, in Latin America, 42%. Mexico has a population of 86.7 million persons. More than 42% are under 15 years of age; only 4% are over 65 years of age. If one goes around the world, one finds that within approximately the next 10 years, a tidal wave of young people will be flooding the world and the questions we ask are: What will they do? How will they gain employment? What sort of jobs will they have?

The crucial question, therefore, is what happens to them? Logically, there are only three things one can do: take their people, buy their goods, or give them capital. All three options are generally unacceptable to different countries.

Japan does not like to take people even though its population is aging. Many countries in the world are trying to send people back. The Germans are trying to send back the Turks who were brought in after World War II as unskilled laborers.

The crucial question, therefore, is what do we do with this accumulation of young people? Some people are moving to the United States; immigration is playing a crucial role in transforming the character of our society. We are taking some goods, by now pushing back the border, the Maquialldora area south of the border in Mexico, to provide new factories. But as a worldwide problem the difficult constraints remain: We take their people, buy their goods, or give them capital.

If one examines a recent U.N. report on Africa, a basket of exports today, compared with 10 years ago, is worth approximately 50%. When you subtract oil, it is worth about 30%. What these figures mean is that with the depression of commodity prices Africa has been locked into primary production. This has been Africa's circumstance for the last decade or more. The inability to move out of areas of primary production locks them into a very difficult economic situation. That situation represents the basic pressure in world economy and world society today.

If one looks at the question of food, we see that overall food production today has risen substantially and the situation which existed until approximately 25 or so years ago, that is, a yearly famine in some part of the world, almost no longer exists.

Except in a few instances, such as Bangladesh, famine in terms of raw hunger has tended to disappear. The situation is one in which food production has become adequate, whereas nutrition has not. Various reports indicate that in developing countries approximately one-third of the population is badly malnourished and exists below even their definition of the poverty line.

My central point, however, is that in almost all instances the food problem is not primarily an agricultural problem. It is not a technological problem. It is largely a problem of politics and socioeconomic organization. And one can take a very dramatic illustration to show this.

Burma was a rice exporting country 25 years ago. In the last 10 or more years it has been a rice importing country. Nothing has changed in the geography. If you fly over large parts of Burma, it is almost like a rice paddy. What happened is a xenophobic dictatorship under Ne Win came into power, closed the country, and the whole economy collapsed. Burma is a perfect illustration of a socioeconomic fact. Ethiopia, subject to famines, always managed because of a small trader system that allowed the distribution of food, subject to certain price increases and, perhaps, profiteering, but, still, a price system. The regime of Mengistu broke the small trader system and larger famines came into play.

If one looks at a geographical map of the Soviet Union, the wheat-growing area from Moscow to Rostov to Lake Bakail is roughly similar to the area of North Dakota and Saskatchewan in the North American continent, although the weather conditions are somewhat more hazardous. Nevertheless, approximately 25% of the population of the Soviet Union is engaged in agriculture, compared with less than 5% in the United States. The collective farms, highly inefficient and highly unable to deal with the question of food, still dominate, and the Soviet Union has become a wheat-importing country.

If one looks at China, a remarkable transformation has occurred in the last 10 years before the tragic events in Tienanmen Square. In China agricultural production grew impressively after one decision: to liberalize a price system for the farmers. From 1980 to 1985 China achieved one of the fastest growing gross domestic product (GDP) rates in the world, exceeding even that of Korea.

The increase in agricultural productivity of almost 10% contributed one-third of China's impressive overall GDP rate. In this situation a change in economic policy was largely responsible for increasing agricultural production in the society. In almost all parts of the world, particularly during the last 20 years, most countries of the world have had somewhat sufficient food supplies. Where supply has not been adequate, the major problems have not been agricultural or technological. The economic policy and the social and political organization of these societies have inhibited production.

On the other side of the ledger there is the question of the surpluses and gluts, largely because of the manner in which the trading systems have been thwarted in many societies by subsidies. Wheat subsidies, for example, in Europe, turned the leading nine members of EC from net importers of 20 million tons of wheat annually in 1965, to net exporters of 10 million tons in 1983. Why should West Europe export wheat? It is uneconomical, but does so for political reason, to protect their farmers.

If one looks at the average value of farm subsidies, the range is approximately 9% of market value and domestic agricultural production in Australia to 72% in Japan (largely of rice). A crucial illustration is that in 1983 the United States imported 200,000 tons of sugar from three countries, that is, Bolivia, Colombia, and Peru. Today, because of import quotas, the United States imports only 85,000 tons. What do you think the South American farmers do? They turn to the production of coca leaves and cocaine, because their sugar markets have already been narrowed by American policy. So, here is a situation wherein economic policies, political policies, protectionism, subsidies, various forms of imbalances, whatever you want to call it, distort basic agricultural production.

The newspapers today are filled with dire warnings and spectacular headlines about the environment. The *Exxon Valdez* affair is the current example, although we have had the question of Chernobyl, that of the ozone hole in the Antarctic, and the question of the greenhouse effect, to name a few.

Part of the difficulty is that the information is unclear. The ozone effect may be a real problem that will worsen, because by killing the algae, the krill are killed, and the entire food chain is disrupted for the fishing industries.

The greenhouse effect may perhaps be a bit more difficult because of cross-cutting situations. As urban densities increase, heat in those areas increases, simply because of the large amount of waste and the large amount of production. We therefore have the question of the balance between urban areas and nonurban areas. Data for the past 50 years are inadequate. The estimates are based on computer simulations. I argue neither for nor against the issue other than to say that media attention often makes it difficult to get the adequate evidence.

There is a broader socioeconomic context in which this environmental issue lies. We have achieved economic growth at an environmental cost. The cost has been the increase in waste (whether it be garbage or hazardous waste) and pollution. Industrial pollution has resulted from our energy systems, that is, fossil fuels, oil, natural gas, and coal. If the pollutants are removed from the air by scrubbers, where do they go? We pollute the rivers. If we do not put them in the rivers, they are buried in the ground and a question of the hazardous waste arises.

One is locked into a cycle basically because of the source of energy supply. We seem to be in a trap. There are two facets that are important.

One question is: How do we allocate the costs of pollution more rationally? One of the chief reasons for the increase in pollution in the post-war years, particularly when economic growth began after World War II, was that air and water were treated as a free good. As the economic textbooks would say, air is a free good. Or if one takes the famous diamond–water paradox of Adam Smith, diamonds are expensive; water is not. Why? Because water is plentiful, and therefore it is cheap.

If something is free, good, and plentiful, you tend to misuse it. There is no cost to you in the way in which you dispose of it. In the 20 years after World War II, agricultural productivity in the United States went up sensationally; in a large way it revamped most of the American farming system. Twenty million people left the farms after World War II because of increases in agricultural productivity.

What also happened was that the chemical fertilizers that increased the productivity ran off into the waters and therefore polluted many rivers and lakes, that is, Lake Erie. The polluters were not charged for the use of a resource. When you begin charging people for the use of a resource, a different pattern of usage may emerge.

The first response has been bureaucratic. Administrative regulations were established to limit the number of effluents one could put into the air. Administrative intervention resulted in higher costs.

We have now moved, perhaps a little more rationally, into market mechanisms to try to manage costs. One method is to have effluent charges, in which charges are levied against firms in terms of the amount of the pollutants discharged. France and Germany levy such charges to raise revenues.

Effluent charges force firms to economize in the kind of pollution they create, but since some firms are more efficient than others, they are penalized by flat charges. A system is being tested whereby for regions as a whole, a level of pollution is established and permits are issued that pay for the pollution. The more efficient can trade those permits to others which are less efficient, although the total level is maintained in terms of the regulation of the pollution itself.

This procedure is a type of control, not a matter of change. Therefore, we need to take a further step, particularly if economic growth is not to be a huge charge on our progress. We must find sources other than the old fossil fuels which create the environmental costs. This search might involve the question of nuclear energy, a very delicate subject. Nuclear energy is pollutant-free, but has other costs in terms of waste and nuclear hazards. Other substitutions must be found. It might be solar, which by and large is still inefficient. Fusion power and superconductivity are clouded possibilities.

My demographic picture is one of imbalance throughout the world, but particularly in regard to youth cohorts, food adequacy, generally a political socioeconomic matter, and the question of the environment.

BACKGROUND REFERENCES

The World Bank. *Social Indicators of Development* 1988. Baltimore: Johns Hopkins University Press, 1988.

The World Bank. *World Development Report 1988*. New York: Oxford University Press, 1988.

The World Bank. *World Development Report 1989*. New York: Oxford University Press, 1989.

World Resource Institute. *World Resources 1988–89*. New York: Basic Books, 1989.

The World Bank. *World Tables 1987*. Fourth Edition, 1987.

CHAPTER 2

Paradigms of American Agriculture

Durward Bateman

A griculture is attracting more attention from nonfarmers today than it ever has in the past. Certainly those of us laboring in the vineyards welcome the expanding interest in the importance of our food and agricultural systems. At the same time we are somewhat apprehensive because the amount of attention and the strength of opinions being expressed frequently seem to be greater than the accuracy of the information possessed by most of the participants in the debates. Agriculture, which has been referred to as the "mother science," has been and continues to be fundamental to human advancement. Even today it is a given fact that without a stable food supply there can be neither peace nor human progress. There is no area of concern of greater importance than agriculture and food, unless it is the matter of the expanding human population itself.

Our modern agriculture and food industries are largely products of the developments in the basic sciences and the application of the resulting technologies. Yet the structure and the status of these industries have been shaped largely by our political and social institutions which have at times facilitated and at times inhibited progress. The land grant universities have served as a primary catalyst for the agricultural transformations over the past century. The colleges of agriculture with their experiment stations and extension services have focused on developing scientifically based agriculture in their respective states and regions. Since the turn of the century science has played an increasing role in expanding the options for agricultural progress. During this period our definition of agriculture has changed, markets have expanded, and the national interest in agricultural progress has shifted from efficient production of food in order to free labor for industrialization toward production of industrial raw materials, value-added products, and international trade.

Over this period of great change in American agriculture there has been a considerable lag between the changes brought about by adoption of new technologies and the social and political recognition of the consequences of those changes. The dominant political rhetoric generally has involved a nostalgia for the past of the Jeffersonian yeoman farmer. Meanwhile, our economic incentives, based on a cheap food policy, have resulted in a highly efficient industrial system of agriculture more aligned with the Hamiltonian perspective. Even today debate over agricultural policy is still shaped by these conflicting ideals.

We have evolved in a relatively short period from a society in which half the population was engaged in a primarily subsistence agriculture to a highly industrialized society in which less than 3% of the population produces, for this nation and for export, the most diverse, largest, and safest food supply in history. From being a labor-intensive industry, agriculture has become our most capital-intensive industry. Our people pay only 12% of their disposable income for food. The miracle represented by this progress has been the foundation for our increased standard of living.

What has happened to farmers in North Carolina reflects what has happened nationally. The only difference is that a greater percentage of the state's 6.4 million population remains in rural areas than has been the case in other states. The 180,000 people in North Carolina still engaged in farming operate some 59,000 farms. Approximately 15,000 of these farms account for nearly 90% of the production. Yet the state has the largest and most diverse agriculture in its history, and it takes place on the fewest cultivated acres in the past 100 years.

The efficiencies of production indicated by this shift in population can be traced through three periods of American agriculture, each represented by a different paradigm. Although there is a chronology implied by associating these paradigms with different historical periods, they are really additive mental models—elements of each persist into the present day. And each model carries with it a different research agenda and additional players in setting that agenda.

In the first paradigm agriculture is farming. The focal unit was the farm, markets were primarily local or regional, and the national interest supported a more productive agriculture, which could release laborers from the land. The research agenda for agriculture as farming is focused on increasing productivity. The focal disciplines were agronomy, animal husbandry, and engineering. The research output was translated into improved plants and animals and enhanced mechanization. The research agenda was set by dialogue among farmers, experiment station scientists, and sometimes state legislators who were often farmers themselves.

In the second paradigm the concept of agriculture shifted from the productivity of the farm to the system of commodity production. Markets became national and international and the national interest expands to include cheap food and foreign trade.

The driving force behind the research agenda was improved efficiency. This brought into focus the disciplines dealing with economics and management and the sciences that contributed to better understanding of the biology of the components of the production system. The output from research now included more efficient management systems, organic chemicals, integrated pest management and better management practices. Those involved in setting the research agenda and funding priorities included environmentalists, agencies of the USDA and NSF, and others, plus critics. Remember *Silent Spring; Hard Tomatoes, Hard Times; The Pound Report;* and the Winrock Conference?

Paradigm three brings us to the present and projects into the future. Agriculture in its modern context is no longer limited to farming nor even to production systems for commodities. We are now beginning to realize that agriculture should be defined as *the interface between human and natural systems.* This view relates agriculture to the management of our natural resource base; it brings into consideration our relationships with the geosphere, the atmosphere, the hydrosphere, and the biosphere, and their management for an efficient, sustainable life-support system. The unit of focus is increased to include value-added products, and the market context is totally international. Our national interests include economic development, international competitiveness, and global environmental impacts.

The focus of the research agenda for this kind of agriculture should form the basis of discussions about the 1990 farm bill. We should be engaged in a great national debate over whether society is prepared to invest in this type of agriculture. What will our future agriculture look like? What will be the role for chemicals? How will we remain competitive in international markets? Will efficiencies gained through new technologies, changes in structure, land ownership, and distribution systems compensate for the additional research and production costs for an environmentally sound, sustainable agricultural and food system? The central disciplines and technologies for a sustainable agriculture will include molecular biology, ecology and the application of expert systems and robotics. The output expands to include a host of new biotechnology products and raw materials for industry. A greatly expanded list of participants is now involved in setting the research agenda.

I see the debate of the next decades focused on the question of whether we will be able to create a profitable system of agriculture that focuses on relationships implied by the concept of agriculture as the interface between natural and human systems. For this to happen the public will need a much broader understanding of agricultural, ecological, and economic issues. Will agriculture be treated as a business in the industrial sector, and will the government develop policies with incentives fostering efficiency and environmental protection simultaneously? This will mean shedding of old ideals and embracing a holistic view of future potential.

I believe that this process has begun, but it will be slow and for some it will be painful. The primary determinants of the future food system will remain the same as those in our recent past: developments in technology fueled by advances in science and a research and policy framework determined by our political and social institutions.

As an administrator of a College of Agriculture and Life Sciences with departments that span the agricultural and biological disciplines, it is not difficult for me to visualize the potential of the future in terms of technologies that could be developed for agriculture or to identify constraints that will affect our future. But I think we are entering an era of paradox. On the one hand, we have unlimited opportunities in science and technology, tremendous public interest in food, and potential environmental crises of global proportions. On the other hand, we have unprecedented levels of public ignorance about agriculture, strenuous social resistance to implementation of technological changes in agriculture, and narrow parochialism on the part of research-funding sources. The interactions of these forces will shape the future structure of our food system.

The advances made in biological and other sciences offer an expanding array of options for developing new technologies to apply to our agricultural and food industries. I do not see that our basic understanding of the undergirding sciences will be a limiting factor in the foreseeable future.

The development of agricultural technologies continues to make rapid progress, although I have some concerns related to the funding and control of agricultural research. In the past agricultural technologies were developed primarily by the land grant colleges. This system was sustained by state and federal formula funding plus an applied research emphasis in USDA/ARS. In recent years we have seen the emphasis shift from federal formula funding to state funding, from applied research to basic research within USDA/ARS, and a greater emphasis on agricultural research by the large agrichemical and seed industries. Furthermore, the major agricultural colleges are dependent for a larger proportion of their support on competitive grants and industry agreements.

These changes have some significant implications. In the past the state–federal agricultural research system served the public interest in the best sense of the term. The current pattern of funding agricultural research tends toward emphasis on short-term parochial interests; development of basic science away from the context of its utilization; increased emphasis on proprietary or protected technology; and deemphasis of coordinated investigations of complex long-term issues.

A major problem in future agricultural progress relates to a lack of general public knowledge of our food production system. This situation is not too surprising since over 97% of our population is "detached" from the land, but it does represent a catastrophic failure of our educational system, which has neglected an area basic to life and to the advancement of human progress.

The general public and many people in positions to make decisions affecting agriculture do not understand the reality of our food production systems. Even the number of farms as reported by the USDA is misleading. Agriculture as an economic sector is concentrated in many fewer units than is generally realized. In most places the small, diversified family farm no longer exists, except in the minds of some politicians, of citizens two or three generations removed from agriculture and in the rhetoric of those who are basically opposed to modern technology. Sectors of agriculture are being transformed into major food industries encompassing all aspects of production, processing and marketing. The poultry industry is essentially vertically integrated. The swine industry and the aquaculture industry are both moving rapidly in this direction. Contract production of fruits, vegetables, and other raw agricultural products by farmers is increasing and will likely increase in the future.

Perceptions having little or no basis in reality can and often do sway public policy. The matters of environmental pollution, food safety, water quality, and so on, all need to be considered in the context of risk assessment and trade-offs in terms of costs and benefits to society. When any of these or other relevant issues are focused on in isolation, it is unlikely that society's best interest will be served. The ability of specialized interest groups to play upon the emotions and ignorance of the public in our modern society is enormous. Coupled with the single-issue approach favored by most advocates, the fear of technology so prevalent today tends to create an environment in which society denies itself useful advances in the food sector.

For example, the use of irradiation for the preservation of food represents a safe technology that has not been accepted. At the same time the use of microwave ovens has readily been embraced. The use of biotechnological products in agriculture appears to be meeting with considerable resistance. For example, the use of BST in the dairy industry, while a perfectly safe technology, may not be implemented because of public fear of employing a biotechnological product in the food chain. Yet the use of products such as genetically engineered growth hormone and insulin appears to be readily accepted in human medicine.

In conclusion, in each case when I consider "solutions" for these contradictions, I am brought back to the need for the public to be better educated and informed on matters related to food and agriculture and the natural resource base. We may indeed be in a race only to see whether or not scientists can understand the dynamics of ecological systems before we alter them irreversibly. But we also may be in a race to see whether or not the public and the policy makers can understand the complexity of the agricultural interface between social and natural systems.

Our political system has great difficulty in facilitating change that requires short-term sacrifices for long-term gains. As George Daniels has pointed out, our model for problem solving since World War II has been the Manhattan Project. The United States, he argues, "has no real drive to support science that is divorced from some presumed crisis." Yet transitions occur though they be slow and more painful than they need be. Large numbers of people have been displaced from the land by changes in technology. The debate over whether or not dairy farmers should use BST to increase milk yields should be a debate about the economic and social consequences of this new technology as opposed to an unfounded fear of its safety.

Despite the roadblocks represented by special interests, contemporary troglodytes, and lack of knowledge on behalf of the public, we shall continue to try to apply to agriculture the best scientific knowledge at our command. I hope there are now enough people in the right places who perceive that the impact our societies' attitudes are imposing on agriculture could lead to a major crisis and that agriculture is so important to our long-term survival that it deserves increased public resources and attention from a more fully informed perspective. I believe that science in the public interest is best conducted in public institutions. They are the most likely to produce objective information relevant to the real world. If we can develop and fund mechanisms for conducting the long-term interdisciplinary studies that will be needed, I am optimistic about the future.

If the agricultural and food industries continue in the directions of the recent past, we will continue to develop a more efficient, productive food sector which will lead to an industry akin to the Hamiltonian ideal. Over time, agriculture and the food industry are merging into one.

BACKGROUND REFERENCES

Bonnen JT. The institutional structures associated with agricultural science: what have we learned? In: Rhodes J. ed, *Agricultural Science Policy in Transition*. Bethesda, Md.: Agricultural Research Institute, 1986:37-70.

Browne WP. *Private Interests, Public Policy, and American Agriculture*. Manhattan, KS: University of Kansas Press, 1988.

Buttel FH, Busch L. The public agricultural research system at the crossroads. *Agric History* 1988:62;303-24.

Daniels G. *Science in American Society*. New York: Knopf, 1971.

Dahlberg KA. *Beyond the Green Revolution*. New York: Plenum Press, 1979.

Dahlberg KA, ed, *New Directions for Agriculture and Agricultural Research*. Totowa NJ: Rowman & Allanheld, 1986.

Mayer A, Mayer J. Agriculture, the island empire. *Daedalus*. 1974 Summer:83–95.

CHAPTER 3

Recent American Agricultural History

George Brown

The economic and technological changes of recent decades have resulted in issues that face us in the corporate, scientific, and policy arenas in the new global community. The extent of these changes can perhaps be perceived by reviewing the last 50 or 100 years in American agriculture. Milk production has tripled over the past 50 years and will probably triple again as a result of the introduction of BST. Similar results have been achieved for the plant sciences using conventional breeding methods.

Eighty percent of the total population in the United States was committed to agriculture 150 years ago; today that figure is down to 2% or 3%. What I perceive happening in the 21st century is continued improvement of productivity. The improvement will have its impact largely in the developing world. That is the issue I wish to call to your attention, because it is going to have a massive global effect that we should understand.

The economic and scientific changes in our society have far outstripped our broader social understanding of the importance of these changes. We have become a global community in which 21st-century nations will become more like boroughs in New York City than the sovereign empires of the 19th and 20th centuries. Domestic, political,and economic decisions must be validated within a global context. Corporate aspirations must be tempered by international conditions. In a world changed by science and technology, researchers can no longer retreat to the value-free neutrality of their ivory tower and abstain from considering the consequences of their work.

I do not propose that every good scientist sacrifice his or her science to become a politician, but a collective concern does have to be expressed in an effective way in regard to the products of our research. Nowhere does this need for changed perspectives become more apparent than in the developing world. Two-thirds of the world's residents live in conditions that would have caused the overthrow of the government if they existed within the boundaries of any developed country. Even without armed revolution the magnitude of the developing world's debt would result in a developed country's economic destruction. And as we become a global community, these conditions will confront those of us in developed countries more than ever before.

Neglecting these problems and the needs and aspirations of the citizens of the developing countries will create global ghettoes from which global, economic, and political disruption will emanate. Just as our own country could not exist half slave and half free, the world cannot exist with one-third wealthy and two-thirds in abject poverty. This problem must be examined and solved long before the end of the 21st century.

At the same time we must recognize that the residents of developing countries represent a human resource, a huge reservoir of potential talent that can contribute immensely to the solution of our common economic problems. Likewise, this large group of people represents the customers of tomorrow, the expanded markets for which the private sector in developed countries will be competing. The competition will be keenest in the food and agriculture sectors, since food is a necessity of life, and since the production, distribution, and consumption of food constitutes the major human economic activity in developing countries. The food and agriculture sector and the policies that support and direct this sector will probably have the most to do with shaping the nature of the global community emerging in the next century.

We will have to learn from our mistakes and not export them. We will have to do a better job internationally than we do domestically in anticipating and planning for the future in a systematic way. We will have to make some short-term domestic sacrifices to achieve long-term international success, something that has proven impossible to accomplish currently with our own farm and economic policy, but which some of our competitors have done with great skill and foresight.

It should be clear from our own roller coaster experience with regard to agricultural exports during the 1970s and 1980s that U.S. agriculture in the 21st century must improve its economic and political intelligence about global trends, improve long-term, long-range strategic planning, and achieve greater stability in the global market. None of this wisdom is new or startling. We have experienced similar dramatic changes in agricultural exports in the past, such as those after World War I; we had similar advice offered and we rejected it, as we generally reject the same advice today.

In addition to the requirements for better strategic planning and improved stability, U.S. agriculture must continue to increase in efficiency and productivity. As Vernon Rutan said several years ago in an article in *Science*, "The capacity of American agriculture to expand its foreign markets and retain its domestic markets depends on continued declines in the real cost of production. American agriculture has achieved its preeminence in the world by substituting knowledge for resources... A necessary condition for U.S. agriculture to retain its status is enhancement of both public and private sector capacity for scientific research and technology development."

A consensus has almost developed that U.S. agriculture and its supporting research and technology infrastructure must undergo major changes if we are to meet the challenges of the 21st century. The nature of these changes is suggested in two recent reports from the National Research Council of the National Academy of Sciences, "Alternative Agriculture" and "Investing in Research."

It should be noted that the U.S. Government today spends approximately $2.1 billion per year on agricultural research at all levels; this amount has changed only slightly in constant dollars in the last 25 years. Private industry (largely the chemical and pharmaceutical industries) has substantially increased investments in food and agriculturally related research so that the amount is approximately equal to government funded research, although directed largely toward more proprietary areas.

The National Research Council's recommendation in their report "Investing in Research" calls for an increase of $500 million per year, or approximately 25%, in government- funded agricultural research over a period of several years. This amount is not out of line with initiatives already approved by President Bush for increases of about 100% in funding for the National Science Foundation over the next 5 years and similar increases for the space program.

The United States spends approximately 3% of its total GNP on research and development, about equally divided between the public and private sectors. The food and agriculture sector spends approximately 0.5% of its portion of the GNP on research and development, an amount totally inadequate to conditions facing us in coming generations.

I emphasize this point because we need to recognize the inadequacies of research funding for agriculture. Reference has been made, for example, to the need for more effort in plant genetic research. The opportunities offered by genetic technology in agriculture, plus the changing nature of agriculture during the next generations require the United States to increase its commitment to the support of research in these areas.

Biotechnological research holds great promise for dramatic change in agriculture in the next century. Euphoria abounds in the discussions of biotechnology and the promise it holds to feed a hungry world. The manipulation of genetic materials to overcome pests and diseases that limit global production is an exciting prospect. Improvements in plant and animal genetics can greatly expand food production. But if any of this promise is to come true, and I believe it will, we must carefully assess the applications of this technology and include in the assessment not just the health and safety aspects, which tend to be the focus of our regulatory activities, but also the political, social, economic, and ethical elements involved.

I am reminded of an earlier advance in agricultural technology, the invention of the mechanical cotton picker, which was not subject to any analysis or assessments of any kind, as far as I know. It greatly increased the efficiency of the cotton industry and it was a great economic boon. Yet, this invention resulted in the largest migration in recent U.S. history, as jobless southern farm workers, mostly Black, moved to urban areas in the North. Without consideration of the social consequences of the increased agricultural efficiency, we haphazardly created a condition of social disruption for which we are still paying over 50 years later.

In hindsight, should we have gone ahead? Probably so, but we should have adequately prepared ourselves and the displaced workers for the ultimate result. This is the kind of situation that faces India, China, and many other parts of the world today as they introduce new agricultural technologies.

Effort must be made to study this problem. The mistakes of the mechanization of cotton picking probably would not be repeated in the United States today because we have a more sophisticated, politically enfranchised population. What will happen, and is happening, is that the affected population is raising the issue of adverse consequences at the start of the new technology's application. If you have any doubt, go to Vermont or Wisconsin and observe the debates on the development and commercial use of BST.

The debate is not really over health and safety; health and safety pose a cover for the real issue, which is displaced dairy farmers. If we fail to develop a strategy to deal with this problem, we will have the same delayed action on every type of new technology in agriculture that affects employment.

Other objections to the use of biotechnology are ethical and religious. There are a lot of people who do not like the idea that a cancer can be put into a mouse because they believe that only God could create cancers and that we should not be doing that sort of thing. We must be willing and able to deal with these attitudes.

Some of you may remember a book written about 35 years ago called *The Technological Society* by a French lawyer and philosopher, Jacques Ellul. In the last chapter of the book which he entitles, "A Look at the Year 2000," he notes the predictions being made by scientists of his time, the mid-1950s: all food will be completely synthetic; the world population will have increased fourfold, but will have been stabilized; disease as well as famine will have been eliminated; voyages to the moon will be common place, as will inhabited artificial satellites; and knowledge will be transmitted directly from electronic data banks to the brain. I am waiting for that day to come about. He goes on to say, "A question no one ever asks, when confronted with the scientific wonders of the future, concerns the interim period. Consider, for example, the problems of automation which will become acute in a very short time. How socially, politically, morally and humanly shall we contrive to get to this future? How are the prodigious economic problems, for example, of unemployment to be solved? ... How shall man be persuaded to accept a radical transformation of his traditional modes of nutrition? How and where shall we relocate a billion and a half persons who today make their living from agriculture and who, in the promised ultra-rapid conversion of the next 40 years, will become completely useless as cultivators of the soil? ... It is not difficult to understand why the scientists and worshipers of technology prefer not to dwell on this, but to leap nimbly across the dull and uninteresting intermediary period and land squarely in the golden age."

Ellul may be regarded as a worrier from 35 years ago, but he may also be regarded as a prophet ahead of his time.

What is still lacking today is a comprehensive policy analysis or technology assessment that examines all the consequences of technological change, or, as with BST, at farm economics and government policies which affect the dairy farmer. I doubt that those involved in developing BST consulted with the affected farmers to see what their needs and opinions were, or what would be required to avoid the potential disruption that we may face. This example of a seemingly good scientific and economic opportunity under attack from many who were to be its beneficiaries can happen internationally unless we do a better job of assessing the technologies we seek to employ to help others climb the economic development ladder in our global community. At a minimum, unplanned technological development can result in economic displacement or loss, as many small dairy farmers in the United States fear. At the worst, inadvertent, major dislocations can result, such as those with the mechanical cotton picker, but on a global scale and with consequences magnified many times over. In fact, the problem already exists in many parts of the world.

Those in the commercial sector might rightly ask if it is their job to solve all of mankind's problems. Of course, the answer is no. But they and the researchers who are working on advances in technology must be ever-mindful of the possible results of their efforts. They owe it to themselves to involve the affected parties at an early stage and to seek the cooperation of the public sector in a combined strategy to minimize any disruptions. This is surely true in the United States and is even more true in the developing two-thirds of the world where the impacts of change will be even greater. Increased sensitivity to this matter will be the best marketing strategy conceived.

This is a new way of operating, a new mindset for all involved. For those of us in developed countries, it involves some giving, or more accurately, sharing of control and power with developing countries and the markets they represent. In the short term this could mean some loss domestically, some change and disruption which must be planned for and dealt with. In the long term these changes will mean increased opportunity and stability that will benefit the entire global community and will benefit those private corporations that handle them well and perceptively.

Perhaps the best observation recently on this situation comes from an unlikely source. In a speech before the Foreign Policy Association in New York, Soviet Foreign Minister Eduard Shevardnadze observed, and I quote, "Radical, bold steps are needed, a kind of new deal, a transition to a policy that would draw the developing nations into the scientific, technological and information revolutions. It will be necessary to overcome a certain psychological barrier to go beyond national concerns and start thinking in global terms."

He draws an analogy between present international conditions and those of the Great Depression. I believe this analogy is correct and believe we need to approach an increasingly interrelated global community with this perspective. This is the landscape in which our corporate sector and foreign policy experts will have to operate well into the 21st century.

Fortunately, at a time when we need to improve conditions globally for our own survival and self-interest, if for no other reason, we have available these powerful technological tools stemming from our scientific research, such as biotechnology. I feel privileged to have been involved at a policy level in the development of this technology. I believe that fate has given me an opportunity to participate in activities that centuries from now will be noted as a major turning point in mankind's progress.

Should the scientist working on these exciting breakthroughs stop his or her work simply because of the social problems described? Should a company working on these breakthroughs be expected to singlehandedly deal with the myriad consequences that they will bring? Should government in developed countries sit back and let developments take their course? The answer to all these questions is no. Each of the sectors involved must work together to anticipate and to minimize any potential disruptions in order to maximize the benefits of biotechnology or any new technology. New modes of thinking will be required, of making the overcoming of psychological barriers, to use Shevardnadze's words, as important as overcoming scientific barriers. I have said before that the ultimate limitation on the development of biotechnology may not be a scientific, but may be a social one. This is not the time for narrow, political ideologies, isolated scientific thinking, or short-term corporate decision making. It is time to take a new look at ourselves and the future we all desire and to move carefully forward. The stakes are too great to do otherwise.

CHAPTER 4

Technological Advances and World Hunger

M.S. Swaminathan

The 21st century, our future, will be entered with fanfare, but we take the present problems with us, and our present problem in the broad area of food and agriculture is the problem of overcoming hunger. According to the latest estimate of the World Food Council over 700 million children, women, and men will go to bed hungry tonight. Today we recognize that hunger is not necessarily caused by food being unavailable in the market. I am not talking in global terms, but mostly in country terms. In my country, India, there is enough food to buy if one has enough money to buy it, so that economic access, rather than physical access, to food has now become the most important food security challenge. Judging from the rate of population growth pressures on land and water and from the extent of biological impoverishment now taking place, ecological access to food might become the most important security challenge in the next century.

I prefer the term nutrition security to food security, because food security, in the FAO sense, is interpreted by political leadership in million tons of grains as reserves. But nutrition security means access, both physical and economic, to balanced diets and safe drinking water to all people at all times. Safe drinking water is important because in developing countries different kinds of intestinal infection result from unsafe drinking water. Even if a child has access to enough calories and proteins, a lot of it is lost because of infection. Hence, provision of clean drinking water should be regarded as an integral part of a national nutrition security system.

Thus, the first problem, whether in the present or next century, is that the extent of hunger is increasing. At the time of the World Food Conference held in Rome in 1974, Henry Kissinger in his inaugural address said that within 10 years we should be able to overcome hunger. At that time the hungry population was only about 300 million. Now this number has increased to about 700 million people and about two-thirds of these people are in Asia.

We live in an age of paradox. On the one hand, we are witnessing spectacular technological advances that result in great opportunities for producing more food. At the same time more people are going to be hungry. In addition to economic access, ecological access to food is likely to become an important problem in the next century in developing countries, largely because the very foundations on which sustainable agricultural development depends are all being eroded. We are witnessing both a steady loss in the biological potential of the soil and in the biological wealth of nations. Soil erosion, water logging, salinity, and different forms of biotic and abiotic stresses are increasing.

Because water resources are not managed scientifically and the use of groundwater cannot be sustained, water-related stresses will increase. The loss of biological diversity is a result of destruction of habitats where biological diversity occurs and various other problems are not entirely under the control of developing countries. Potential changes in climate and sea levels will further erode the capacity of developing countries to bring about sustainable advances in biological productivity. Gene loss erodes the potential to derive benefit from genetic engineering.

The nature of the challenge can be seen today in China, where food has to be produced for over 1 billion people from a per capita availability of land of 0.1 hectare or 0.25 acre. In India today the per capita availability of land for agriculture is 0.15 hectare. Obviously, the per capita amount will decrease as we reach a population of 1 billion in another 15 years.

The imperative is to produce more and more food from less and less land. It is not only food, but also a variety of other agricultural products, including export commodities, that must be produced from less land. Therefore, ecological sustainability is an absolute must in developing countries that are population-rich but land-poor. For this purpose we need a sustainable agricultural matrix. Such a matrix helps to achieve harmony between the packages of technology, services, and public policies introduced and ecological, equitable, economic, and efficient imperatives.

The second important difference that is growing between the developed and developing countries is the nature of technology required. Our farms are small and they will get smaller still. In animal husbandry, for example, we have in India nearly 22–23% of the world's cattle and buffalo population, nearly 25% of the world's goat population, and 22 % of the sheep population. Nevertheless, the total amount of land we have for grazing and forage is about 3.5–3.7 % of all available land. You may wonder how 25% of the world's farm animal population can be raised with very little grazing land. The answer is that our entire strategy of animal husbandry is based on stall feeding. To our farmers, mixed farming, which involves crops and livestock is both a way of life and a means to security of livelihood. There is a symbiotic relationship between the crops and animals. Ruminating animals are generally preferred. Our strategy to provide some income to landless labor families is to give them animals. In your case animal farming is a land-requiring industry. In our country it is landless labor who are given animals. In other words, technologists, particularly biotechnologists in your country, should take note of this basic difference in methods of raising animals.

Three issues must be raised as developed and developing countries meet biotechnological advances. First, there is an enormous need for a rapid technological upgrading of agriculture in developing countries, particularly in countries such as India and China, where agriculture has to provide not only more food, but also more income and more jobs. Without technological upgrading there is no possibility of producing more food on a sustainable basis. Crop varieties are being bred that possess tolerance to drought, salinity, and pests. Genetic engineering offers the potential to restructure the morphology and physiology of plants and animals.

If these advances are needed in developed countries, they are even more urgently needed in developing countries where the triple alliance of weeds, pests, and pathogens causes greater threat to crop security because of year-round cropping.

We also face a wide range of biotic and abiotic stresses that enlarge the risks faced by farmers. A majority of farmers in the developing world lack adequate resources for strengthening the infrastructure necessary for high yields. Their decisions on land and input use are influenced by the cost, risk, and return structure of farming. Thus, we need land-saving agricultural technologies and grain-saving animal husbandry. We need agricultural practices that are low in cost, but high in yield.

I hope some of the emerging technologies will improve the productivity, profitability, stability, and sustainability of small farm agriculture. Biotechnology, space technology, information technology, and microelectronics can help upgrade small scale agriculture. Remote-sensing techniques help survey and monitor natural resources.

Weather satellites certainly have added a new opportunity for short-term forecasting. In developing countries a marriage is needed between survival techniques and sophisticated technologies.

Traditional technologies should be retained in several areas, largely because several of them have ecological and socioeconomic strengths and are employment-intensive. Until we have alternative sources of employment, it is important that agriculture remains labor-intensive. Job destruction and job creation must be concurrent events if we are to avoid hardship and increased hunger.

The second issue is the urgent need for technologies for the future. Such technologies must lead to the upgrading of skills in rural areas, particularly for women and others who are either underpaid or unpaid. Unless women have independent access to income, it is very difficult in poor families to increase the household income and to improve child nutrition. That is why generation of employment has to become an explicit research objective. Developing countries have opportunities for employment, but rural people often lack employable skills. Illiteracy is not an impediment to teaching new skills; the skills can be learned by doing. An integrated approach to on-farm and off-farm employment is essential to reduce the number of people engaged in routine operations of farming.

The opportunities are great. Certain aspects of biotechnology have become important. Tissue-culture techniques, particularly for the multiplication of superior clones of forest tree species, the whole area of bioprocessing, biomass utilization, microbiological enrichment of wastes, and the chemical engineering industry as applied to biomass utilization, all can help in the preparation of value-added products from the entire biomass. Biomass derived from plants and animals is the principal industrial feedstock available in the rural areas of most developing countries. Biomass is often used, however, in a manner that does not provide full benefit. Using cellulosic material we can prepare balanced diets for stall-fed farm animals. Consider the development of biomass refineries.

This takes me to the third issue; that is, the growing divergence in opinion and perceptions and the emergence of conflicts between developed and developing countries in agriculture. I must stress, however, that agricultural exports are the major means for most developing countries suffering from debt burdens to earn foreign exchange. Therefore, developed countries should take a long-term view on the opportunities. They should help create export opportunities for developing countries. Obviously, developing countries cannot catch up in sophisticated technologic advances, and if they cannot export value-added agricultural products, there is no means at all by which to recover from the debt burden.

It is important that we give serious thought to promoting the complementary evolution of agriculture, a complementary evolution in which both the 4 billion people of the developing world and the 1 billion people of industrialized countries can derive economic and social benefits from the technological upgrading of agriculture.

One area of controversy that has been aggravated by the potential to move genes across sexual barriers by genetic engineering relates to the commercial exploitation of biological diversity. Much of the world's biological diversity exists in the developing world. Tropical rain forests alone are believed to have nearly two-thirds of the world's species.

Controversy has emerged over the patenting of new breeds of crops and animals. If, for example, a molecular biologist or a conventional breeder adds a new gene to a Murrah buffalo from Pakistan or India, the question arises, "What is the contribution of those who developed the Murrah buffalo and what is the contribution of the scientist who has added one gene?" How can the farmers who developed the original material be compensated? This issue is often referred to as "Breeders' Rights" and "Farmers' Rights" and is being discussed in international forums such as FAO. I hope such questions are resolved so that biological diversity is made available for human benefit everywhere. To be sustainable, development must be equitable. I would suggest that formal and informal innovation should both receive concurrent attention and reward.

Nevertheless, as intellectual property rights and patent rights expand, we should consider how the people who go to bed hungry today can derive benefit from socially relevant new technologies. The green revolution spread fast because the technologies were the products of the public sector or philanthropic foundations' enterprise. The gene revolution is becoming increasingly a private sector enterprise. Will the benefits of the gene revolution be available to all farmers in the same way as the benefits of the green revolution?

When we deal with food, which is first among the hierarchical needs of human beings, it is important to consider the kinds of methodologies that must be adopted so that the latest advances in technology become available for human benefit everywhere.

The United Nations Environment Program and the International Union for the Conservation of Nature and Natural Resources are working on a global convention on biological diversity. We must develop methods of conserving, evaluating, and utilizing biological diversity for the common benefit of humankind.

Sharing of genetic resources and technological benefits will be increasingly important topics for international dialogues. In genetic engineering we have an opportunity to help the economically handicapped producers and consumers. The question of safety has arisen, for example., food irradiation. In areas such as food irradiation we need to develop methods by which to assess wholesomeness under conditions in which undernutrition and malnutrition prevail. What kind of safety studies should be undertaken? Can we recommend a food product just because it has been cleared by the FDA or is there a need for nutritional studies based on a different set of parameters? A whole series of questions of this kind arises when we develop criteria for assessing the wholesomeness of food, whether it is irradiated food or products that have been genetically engineered.

We must develop more and more methods of understanding the strengths and weaknesses of tropical and subtropical agriculture. We must develop methods by which the advances of modern science and technology can be shared with developing countries, because the gap is very wide, and grows wider every day, in terms of development and dissemination of advanced technologies.

Industry should take the lead in developing methods to make hunger a problem of the past. President Roosevelt, when he helped to establish the FAO in 1944, mentioned that the world has the capacity to overcome hunger, but this goal has yet to be achieved.

We have the capacity to overcome famine because famine evokes immediate response, but chronic hunger is a deep-seated problem. Ending chronic hunger requires concurrent attention in both the fields of production and distribution. Governments and industry have a joint responsibility in this task.

PART II

THE BIOLOGICAL REVOLUTION, AGRICULTURE, AND THE FOOD INDUSTRY

ABSTRACT

While advances in plant and animal genetics hold out new promise for feeding the world's growing population, public mistrust of the new technology stands as a formidable obstacle. From the Minotaur, a mythological monster—half-man, half-bull—who devours scores of Athenian youths, to Mary Shelley's Frankenstein, popular literature has given genetic engineering a bad image.

Donald Glaser, a molecular biologist and Nobel laureate, points out that advances in genetics over the last generation have successfully pinpointed genes responsible for specific characteristics in plants and animals and manipulated those genes to change the organism and its offspring. These techniques, called genetic engineering, enable us now to make predictable, rather than random, changes in living things for agricultural purposes as well as for medical and industrial purposes. Our scientific abilities are advancing so rapidly that they require economic and public policy decisions on how to use them. For example, as Nobel laureate Arthur Kornberg points out, the gene responsible for insulin production in human beings can be inserted into a bacterium, turning a bacterial culture into a factory for the production of human insulin.

The implications for agriculture are enormous. By implanting genes that produce natural pesticides, tomatoes and potatoes can be made resistant to caterpillars. Soybeans and corn, each of which are relatively deficient in one amino acid, may be engineered to become more complete sources of protein. Biotechnology may even succeed in putting flavor back into supermarket tomatoes by inserting a gene that will allow producers to ripen them on the vine. (Currently, vine-ripened tomatoes are considered too soft for shipping, so they are picked green, refrigerated, and then gassed with ethylene to bring on the red color.) Plant biotechnologist Winston Brill predicts that in two or three years, scientists will be able to apply genetic engineering to all of the major crop plants.

Unusually dramatic advances are occurring in animal genetics. In 1988, Harvard University was granted the first patent for a genetically engineered animal, a tumor-prone mouse to be used as an animal model for studying breast cancer. The creator of the so-called Harvard Mouse, Professor Philip Leder, outlines how genetic engineering has already begun to revolutionize the food industry. In addition to genetically engineered growth hormone, we can also look forward to leaner pigs, chickens and steer, and to fish that reach a marketable size much more quickly. Moreover, the implantation of genes that carry immunity to common viral and bacterial diseases will make these new animals resistant to disease.

From the public's fears about tampering with nature, to the small farmer's resistance to paying royalties for the use of genetically engineered animals, the field of biotechnology is fraught with controversy. All the contributions to Part II emphasize the need for better communication with the public in order to answer these concerns.

CHAPTER 5

Introduction: Biotechnology and Industry

Donald Glaser

The biotechnology industry exists today because of very wise and generous investments by the National Institutes of Health and the National Science Foundation over 40 years. Early in that period, a group of physicists, chemists, and others joined in a campaign dedicated to exploiting *Escherichias*. *E. coli* is a common intestinal bacterium, essentially the simplest free-living organism. The goal was to determine whether, at least in the case of this organism, the laws of physics and chemistry are sufficient to explain everything about the organism.

It was a latter day scientific attack on vitalism. Out of that effort came a wealth of scientific riches, the most important of which was the discovery that one molecule, the DNA molecule, provided the mechanism for inheritance in all living things except for certain viruses that use RNA, a close relative of DNA, as the genetic material.

The discovery that all living things depend on exactly the same molecule for their genetic function meant in principle that we could hope some day to transfer genes from one organism to another. In this way an organism could "learn" to make products that it did not produce normally. All of this was good science. Science, of course, is an activity in which a goal is stated, but in which the outcome is totally uncertain. No scientist would really be excited by a problem for which the result was certain.

We now rather arrogantly use the term genetic engineering when molecular biology is applied to commercially important problems. This expresses a rather high level of arrogance because engineers claim they can produce a particular product in a given time at a given cost. That is almost true of genetic engineering. We can now transfer a human insulin gene into a bacterium and do a certain number of technical tricks, and guarantee that the bacterium will produce human insulin. It will not produce hemoglobin, it will not produce fingernails. It will surely produce insulin. One can guarantee that outcome.

One cannot guarantee that the level of production of the insulin will be sufficient for commercial success, but we can guarantee that it will be insulin and that it will have the proper effect if administered as a drug to humans. We are now in a position to accept goals given to us by physicians or by industrial engineers asking for this or that product, and we can guarantee to make the desired product. Both the development time and ultimate cost of production can be roughly estimated.

Several remarkable techniques have been developed and great advances have been made in the speed of developing a new product and to testing its efficacy. Such technology will continue to improve and will allow us a wide choice of possible products and processes.

CHAPTER 6

A Primer on the Molecular Biology of DNA: Its Modification and Role in Heredity

Arthur Kornberg

If we begin with a sperm and an egg of a given species, an individual emerges with the characteristics of a human or a mouse. DNA is the chemical substance in the genes and chromosomes responsible for what we are. Rearranging the DNA produces novel genes and organisms with novel properties.

The chromosomes are a construction manual for making the cell or organism. This construction manual has 10,000 or more genes, each of which possesses one of the many thousands of instructions needed to make up a particular kind of cell and organism. To understand the dimensions of a gene, I like the miniaturization analogy first suggested by the late and eminent physicist, Richard Feynman. He offered a prize some 20-plus years ago to anyone who could miniaturize the 24-volume *Encyclopedia Britannica* to the size of a pinhead. This prize was actually claimed a few years ago by a graduate student at Stanford. Yet this miniaturization is not the ultimate, in fact far from it. The dot at the end of a sentence of that miniaturized *Encyclopedia Britannica* covers an area which can include a thousand atoms. So we see that the ultimate in miniaturization is atomic language. This is what nature attained over a billion years ago in DNA, the atomic language in which genes are written.

What we have learned about DNA has come initially from the viruses that infect bacteria. A virus of medium size attaches to the surface of a bacterium and injects its DNA, thus starting a cycle of infection. The DNA, the chromosome of the virus, can be seen in the electron microscope when exploded out of the viral head. It is like a fiber, a linear array of 100 or so genes. We know their arrangement precisely; each gene encodes a particular protein.

The bacterium infected by this virus is the common intestinal inhabitant, *Escherichia coli.* Its chromosome is hundreds of times larger than that of the virus and contains 4,000 genes. Each human chromosome is far larger than the bacterial chromosome, and we have 46 of them.

DNA has two major functions. One is to be transcribed into the similar language of RNA (ribonucleic acid), which carries a message to be translated into proteins. The proteins are the machinery that permits a cell to develop into whatever it is going to be and carry out its unique functions. The second function of DNA is to serve as a template to be replicated into an additional copy of the DNA so that when the cell divides into two daughter cells, each will have a construction manual identical to that of its parent.

To complicate matters, some tumor viruses and the AIDS virus posses RNA rather than DNA. When the viral RNA enters a susceptible cell, it is transcribed (in reverse) into DNA and enters the chromosome of the host cell. There it can either be replicated along with the host chromosome and on occasion induce a tumor or some other kind of trouble. Or, it may be converted back to RNA to become a virus again. Everywhere in nature, information flows from a nucleic acid (DNA or RNA) to generate proteins, never the reverse.

DNA is a double helix, a familiar icon. It is spelled in a language of four letters: C, A, G, and T. The virtue of this double-stranded structure, pointed out by Watson and Crick in 1953, was that it provided the correct model for its duplication. The structure of DNA and its replication depends on the very simple fact that C in one strand is bound to G in the other; A in one strand is bound to T in the other. That complementarity of A to T and G to C is the most basic rule for all living things.

To duplicate the DNA strands in order to make copies for daughter cells, the two parental strands are parted and for every C, a G building block is put in place, for every G, a C; for every A, a T; for every T, an A. It can be seen at once that the two daughter molecules have the identical sequences and each is identical to the parent.

In 1955, we found an enzyme in cell juices that assembles the C, G, A, and T building blocks into long chains in the same sequence provided by a preexisting DNA template. How precisely? After some 12 years of trying, we finally were able to demonstrate that our enzyme could make the chromosome of a virus that infects *E. coli.* It was fully infectious and so its 5,386 units (Cs, Gs, As, and Ts) must have been assembled without error.

This experiment created a stir because it appeared to the media that we had "created life in a test tube." President Lyndon Johnson, when made aware of it, congratulated us and the enzyme (DNA polymerase). "It opens a wide door to new discoveries in fighting disease and building healthier lives for mankind."

Nevertheless, we were chagrined that in this synthesis a fragment of DNA was needed to start the synthesis; it became clear that our DNA polymerase could not start a chain. Not only were we unclear about this first step, but other findings informed us of many other complexities in the replication process. By focusing on the tiny, simple chromosomes of the small viruses that infect *E. coli*, we began to observe the wheels within wheels of the replication machinery. It was far more complicated than we had imagined. More than 20 working parts are needed to start a chain and copy a template. DNA is copied at a rate of 1,000 units a second; errors are less frequent than 1 in 10 million. In our current research, we focus on the elements of this machinery, how they do the elegant job of advancing replication at breakneck speed with extraordinary fidelity.

Emerging from the pursuit of a curiosity about DNA, supported for over 20 years by the National Institutes of Health, we learned how DNA could be taken apart and reassembled. As a result, we could contribute reagents and technology for making recombinant DNA. The most recent addition to this technology, perfected at the Cetus Corporation, is PCR, the polymerase chain reaction, which makes it easy to find a gene as remote as a needle in a haystack.

With regard to recombinant DNA, a key discovery was that bacteria have an array of scissorlike enzymes that cleave specifically at certain places. For example, one such enzyme in its patrol of a chromosome cleaves to DNA whenever it sees GATC on one strand, matched, of course, by CTAG on the opposite strand. Staggered breaks provide a lap joint for rejoining by a sealing enzyme. Like all enzymes that act on DNA, it does not matter whether the DNA comes from a bacterium or a human.

When two cleaved DNAs, one from a bacterial and the other from an animal chromosome, are mixed, they will unite at these lap joints and generate recombinant DNAs. In this manner, a segment of DNA from an animal chromosome can be inserted into a bacterial chromosome. For ease of manipulation, tiny pieces of a bacterial chromosome, called plasmids, are employed. They are small DNA circles one thousandth the size of the *E. coli* chromosome. The plasmid is marked with a gene that destroys an antibiotic and therefore confers resistance to that antibiotic on any bacterium that harbors the plasmid. Out of a huge population of bacteria, a billion or more, it is easy to select the one that has a recombinant DNA plasmid by virtue of the antibotic- resistance marker and the unique genetic feature obtained from an animal or other chromosome. This bacterium can then be grown into huge masses in which each bacterial cell produces the substance encoded by the foreign gene in the recombinant DNA plasmid. In this way, bacterial factories for human insulin, bovine somatotropin (BST), and other precious hormones and vaccines have been built.

BACKGROUND REFERENCES

Alberts B. *Molecular Biology of the Cell*. Second edition. New York: Garland Publishing, 1989.

Darnell J. *Molecular Cell Biology*. Second edition. New York: WH Freeman Co., 1990.

Watson JD. et al. *Molecular Biology of the Gene*. Fourth edition. Menlo Park CA: Benjamin-Cummings Publishing Co., 1987.

CHAPTER 7

Plant Genetic Engineering and the Food Industry

Winston Brill

Plant genetic engineering has tremendous potential in the global food system. The technology will improve our health and the environment and will open new markets. As with all new technologies, some markets will be replaced. As standard breeding has not produced Andromeda-type organisms, genetic engineering likewise should be safe.

Genetic engineering techniques have not yet produced commercially valuable plants. Most elite lines of important crop plants have not been amenable to the technology. This situation, however, is beginning to change. I predict that most important crop plants will be routinely engineered within the next 3 years. Thus, investor and corporate interests should increase as a result of the shorter time to produce improved and new plant products arising from applications of genetic engineering.

Most plant genetic engineering activities have been centered on producing plants that resist specific pests. The hope is that fewer dangerous chemical pesticides will be used. We know that unwanted chemicals get into our food supply and drinking water. We know also that farmers do not like to handle chemicals. Chemical pesticides also kill beneficial organisms that destroy other pests. In addition, it is just a matter of time until the pest becomes resistant to the chemical.

Tomatoes and potatoes have been engineered to resist caterpillars, viruses, and certain bacterial pathogens. These resistances are specific for the target pest and are carried through the seed from generation to generation. There is no toxicity to man. Thus, farmers should save money by using fewer chemical pesticides. There is reason to believe that the pest will encounter far more difficultly overcoming the killing activity of the plant. These plants, if used widely, should make agriculture more environmentally compatible than it is currently.

As the technology advances, and it is advancing rapidly, ideas are being stimulated. For example, a tree in West Africa has berries that contain a protein, thaumatin, that is many thousands of times sweeter than sugar. For every protein, there's a gene, and the gene for thaumatin has been isolated. If this gene is incorporated into the chromosome of grapefruit, the resulting fruit will be sweeter. Perhaps as a consequence, the market for grapefruit would be dramatically increased.

Until now, the grapefruit breeder was only concerned about plants that could breed with grapefruit—citrus plants. Now, however, breeders should have much wider choices, because genes from any organism are potentially valuable to the target organism. This single example is yet another argument for maintaining the world's germplasm. Each organism has hundreds of thousands of unique genes. I believe that some of the most exciting work in this area in the next decades will be learning how to search the pool of genes to find those useful to agriculture.

One concern that has to be resolved is ownership of the genes. For instance, several developing countries are frustrated that developed countries freely exploit a gene from a plant indigenous to the developing country. Does the developing country "own" the gene?

Wheat, of course, is an important crop. Laboratories are isolating the genes required for the proteins that alter flour quality. Over the next decade, varieties of wheat will be optimally tailored for the specific type of flour required.

Researchers are attempting to understand the molecular biology of drought tolerance, which is particularly important to certain developing countries. For instance, if a crop can survive for a week or two more than it normally would through a drought, its survival could make the difference between starvation and survival for many people.

There are attempts to apply genetic engineering techniques to plants that will enable them to utilize fertilizers more efficiently. Much of the pollution in our lakes and streams results from fertilizer wash-off after a rain storm. This type of development should save money for the farmer.

We hear predictions that the world's climate is going to change over the next few decades. If this is true, then agriculture will certainly be affected, and the damage could be considerable. Although genetic engineering cannot replace breeding, genetic engineering techniques may be critical because a new line of plants can be developed through genetic engineering much more quickly than a new variety can be developed through breeding. This time factor could be crucial.

Soybeans are a valuable crop. The amino acid, methionine, must be added to soybean meal to make it nutritionally balanced. Thus, poultry feed and soy-based infant formulas require methionine. Laboratories have identified proteins that are high in methionine, and they are being introduced into soybeans. The objective is that the resulting soybean be a more nutritionally balanced food.

At least one laboratory is using genetic engineering to try to remove the major off-flavor of soybean meal. Success could be important to the food industry.

Scientists are looking at oil production pathways and see potential, for instance, in using one type of soybean to make a desirable cooking oil and another to make a good salad oil. Another possibility would be to engineer soybeans to produce omega-3 fatty acids, those found in fish. Such oils are currently recommended as health promoting oils. Thus, different lines of soybeans could be matched to specific market needs.

Research activities are underway to engineer a plant to produce industrially useful products, for example, potato-producing enzymes useful to the food industry. There is even research in progress to have precursors to plastic produced in the potato. Thus, future plastics may come from renewable resources, rather than from petrochemicals.

Plants are an important source of coloring and flavoring agents. The production of such plants may be increased through genetic engineering. Another possibility is to incorporate the genes for such chemicals into crops that are easier to grow in the countries that have large markets for them. It may be possible to totally bypass the field-grown plant by growing cultures of plant cells in fermenters to produce the colors or flavors. In fact, the Japanese have had some success in the production of shikonin, a coloring agent, by fermentation of cells from the plants that normally produce the material. This advance could result in losses of important markets in tropical countries that supply much of the plant material from which such chemicals are extracted.

Plastic wrapping material is important to the food industry. Most plastic that we use today can last 200–400 years in a landfill and is causing a variety of environmental problems. However, a bag has been developed that is made up of polyethylene and cornstarch. This bag is degraded in 6 months to 2 years. So far, the cost of such a bag is relatively expensive. If the granules of starch in corn had different properties, the cost of the bag should decrease. There is interest in using genetic engineering to alter corn so that the starch granules will be more useful for production of biodegradable plastics.

I believe that the new biotechnologic advances will actually "bring us back to nature," a shift that may be necessary for this large population to live on this small planet. Because advances are rapid, however, a major problem in many developed countries is public perception. The public has a poor understanding of the technology. Groups, for their own purpose, have used the lack of understanding of the public to inflame fear of genetic engineering.

Many scientific organizations, however, have discussed safety and environmental issues related to genetically engineered organisms. For example, the

U.S. National Academy of Sciences has two reports to indicate that a genetically engineered organism should be no more dangerous than that same organism genetically altered by traditional mutation or breeding.

Actually, a genetically engineered organism should be considered less dangerous. For example, if the thaumatin gene is incorporated into grapefruit, the resulting new line will be more understood than a new variety of grapefruit resulting from breeding, which is a random mixing of hundreds of thousands of genes. In the case of thaumatin, only a well-understood single gene has been added to the chromosome.

A backlash has occurred, however, to genetically engineered products. Many supermarkets have now banned milk from cows injected with bovine somatotropin. It is extremely difficult to perform a field test of a genetically engineered microorganism. Some communities have banned such tests. It is less difficult, but still quite expensive, to perform a field test of a genetically engineered plant. The onerous regulations are producing a marked disincentive to researchers, especially academic researchers, to develop potentially valuable genetically engineered products for agriculture.

Technical barriers are being overcome. Public perception and regulatory problems will hopefully diminish, and plant genetic engineering will become a valuable technology that will produce food more efficiently and will aid the environment. Commerce will surely be affected, as will the food industry. The potential of plant genetic engineering is tremendous!

BACKGROUND REFERENCES

Flavell RB, Goldsbrough AP, Robert LS, Schnick D, Thompson RD. Genetic variation in wheat HMW glutenin subunits and lthe molecular basis of bread-making quality. *Bio/Technology* 1989;7:1281–5.

Gasser CS, Fraley RT. Genetically engineering plants for crop improvement. *Science* 1989;244:1293–9.

Lawson C, Kaniewski W, Haley L, Rozman R, Newell C, Sanders P, Turner NE. Engineering resistance to mixed virus infection in a Commercial potato cultivar: resistance to potato virus X and potato virus Y in transgenic Russet Burbank. *Bio/Technology* 1990;8:127–34.

Mazur BJ, Falco SC. The development of herbicide resistant crops. *Ann Rev Plant Physiol Plant Molec Biol* 1989;40:441–470.

McCabe D, Swain WF, Martinell BJ, Christou P. Stable transformation of soybean (Glycine max) by particle acceleration. *Bio/Technology* 1988;6:923–6.

Vaeck M, Reynaerts A, Hofte H, Hansens S, DeBeuckeleer M, Dean C, Zabean M, Van Montagu M, Leemans J. Transgenic plant protected from insect attack. *Nature* 1987;327:33–7.

Witty M. Thaumatin II—a palatability protein. *Trends Biotechnol* 1990;8:113–6.

CHAPTER 8

Genetically Engineered Animals

Philip Leder

A lthough global predictions are virtually oxymoronic, those of us who look at a small sector of the scientific spectrum, a small sector of the economy, can make certain predictions that will be valuable, particularly over a period of 4 or 5 years. On the basis of what we do now know, it seems that certain developments will be almost inevitable.

In fact, I can think of few areas of our future existence that will be more importantly and seriously affected by technological developments that are taking place today than the area of food supply and production and its sister biologic field, the area of medical care and diagnosis. Anyone who reads the popular press knows that for the last two decades a remarkable revolution has been ongoing in the field of genetics. Not only has this revolution delivered to us an enormous wealth of fundamental scientific insight that will accrue to our benefit, but it has also provided an array of genetic tools that are extremely powerful in their ability to be pressed into the practical service of mankind.

Having have been involved literally day to day in genetics for over the last 25 years, I continue to feel the pulse and excitement of a field that moves at a pace one could hardly have imagined three decades ago. This pace has implications for the real world, and those implications could not have been imagined as recently as 15 years ago.

I shall outline some of the principles and methods that I believe will have an impact on the manner in which we provide food and fiber from animal sources. In addition, I shall outline some of the achievable short-term goals, some of the technological barriers, and indicate the prospects for overcoming them. I shall also discuss the time for these advances and will mention some of the nonscientific factors that will continue to influence the introduction of this technology into the business of providing food for a growing global population.

I think it is clear to all of us that generations of farmers have understood the general, if not the scientific, principles of animal breeding. Certain characteristics are judged for a variety of reasons to be favorable. Breeding animals with these characteristics tends to produce animals with the same desirable characteristics; over generations, these characteristics become stabilized in a breed.

It is not without reason that the Golden Retriever is such a lovable, slobbering dog, and the Doberman Pincher quite a different beast. By the same token, there are fast racehorses, and there are the others. Heritable differences, and the ability to identify the specific genetic determinants of those behaviors and other properties of animals are issues that have concerned farmers and animal breeders for thousands of years. We can hope to have some scientific basis for identifying these determinants, even those as elusive as must be involved in the breeding of a fast racehorse, and begin to think about the applications and implications that this knowledge can have.

The principle is true for cattle, sheep, hogs, racehorses, birds, fish, and oysters. Obviously the laws of genetics apply to all forms of life.

In the time since Mendel—that is, during the 20th Century (after the rediscovery of Mendel)—breeding attempts have become far more sophisticated, and slowly over nearly a century, we have seen the emergence of high-producing dairy cattle. Indeed, classical cattle breeding and improved nutrition have raised the average milk yield of a dairy cow in the United States from 4,600 pounds per cow per year in 1941, to approximately 14,000 pounds per cow per year today.

The possibility exists of raising milk production to approximately 50,000 pounds per year per cow. That development, depending on whether it is achieved with somatotropin, or whether it is done with genetically engineered animals, could occur over the course of the next 5–25 years. The potential impact and time scale involved begin to emerge.

There are two basic techniques that will bring about the revolution of animal husbandry in the agribusiness. One will result from the use of the products of recombinant DNA technology, somatotropin or viral vaccines, for example, but others will result directly from genetic engineering of animals, and that is the technology I will discuss.

How can we genetically engineer animals? How difficult is this technology? Is it something that a high school student can do? What are the limitations? How can we produce what we have come to call transgenic animals, animals that carry genes that have been introduced to them using the techniques of molecular genetics?

My own ranching experience comes from running a herd of some 5,000–

7,000 animals. These animals are mice, not dairy cattle or pigs, but the principle and the possibilities are similar. Only the space involved and the generation time and litter size are different.

We wished to create an animal model for human breast cancer, a mouse model that could be used in an effort to understand a disease that affects over 140,000 American women each year and in an effort to develop more effective treatments and diagnostic techniques.

One of the great barriers to the development of effective therapies and diagnostic tools is the lack of adequate animal model systems. The availability of these models will greatly facilitate the development of treatments for cancer and other diseases. The techniques we have used for that medical purpose are similar to those we would use to produce a high-growth/low-fat steer. Only the time involved and the venue, obviously, would have been different.

The first step in creating a transgenic animal is to isolate a gene that you believe is important. If you wish to breed a Doberman Pincher with a characteristic of altruism, despite this facetious suggestion, a gene for altruism must be isolated. The gene must be identified that will convey with it desirable properties.

We used a gene that we had identified as being involved in the development of malignancy, a so-called oncogene, a cancer- causing gene. This gene was to be introduced into the ova and the sperm cells of a mouse. That animal would then carry the genetic propensity to develop a malignancy.

Although we began with a mouse, it could be a racehorse or a hog or a cow or a sheep or a goat. The animal is allowed to mate about 12 hours before her fertilized eggs are harvested. The eggs are then teased apart and, using a needle, we inject about 500 copies of the gene that we are interested in into the nucleus of the fertilized egg. The fertilized egg is then incubated for a short time and transferred to the uterus of a pseudopregnant foster mother that would have mated the night before with a vasectomized male.

A proportion of the eggs will take up the injected DNA. The process that governs the uptake, which we cannot control, is one of the variables that must be understood as we develop this technology. For example, we do not have the ability at present to replace one gene with another. We do not have the capacity to replace a good gene with a better gene. Such technology is being developed and should be available within 3 or 4 years.

At present, we can only add a gene. If, for example, we were concerned with modifying the wool structure of sheep, we could add one of the keratin genes that is involved in producing wool, modify it, and have an additional gene. If, however, we wanted to create a more silklike texture by making modifications in the structure of a gene, that must be done with a replacement technology, as yet to be developed.

In regard to the rats, approximately 10–30% of these treated eggs will have taken up the gene we added. Actually, you nip off a little bit of the tail and sensitive techniques detect the fact that the foreign gene has been taken up.

The costs involved in setting up a laboratory to execute this technique are not great. A moderate-sized laboratory for this technology using rodents can be set up and can be operated at an annual cost of several million dollars. If there is a specific project involved, there would have to be good molecular genetic backup facilities to identify the specific genes of interest. If you are using larger animals, the cost would be commensurately greater, because of small litter size and because of generation time differences. A mouse reaches sexual maturity in three or four weeks. The gestation period in a mouse is three weeks, as opposed to nine months for a cow, and dramatically influences the pace of accomplishment.

We have bred transgenic mice that develop small tumors, that is, adenocarcinoma of the breast. The mice transmit to their progeny that carry this gene the propensity to develop adenocarcinoma. We have created a cancer-prone family.

Experiments have been conducted in which growth hormone genes have been introduced into swine. The results suggest that the introduction of this particular gene, although it has no adverse effects on the physiology of the rodent, does have certain adverse effects on the physiology and fertility characteristics of swine. An animal has been produced that more efficiently converts feed into protein, and the ratio between fat and meat on the carcass has been fairly dramatically altered. Pork chops from the identical rib of a transgenic animal and a control animal are different in appearance; the transgenic pork chop is very lean and less marbled than the one derived from a nontransgenic litter mate. Thus, the implications of this work begin to emerge. We can see the possibility of not administering somatotropin, but, in fact, introducing the gene for somatotropin that could be turned on for a short period of time and then turned off.

As the technology improves, and as advantages become more compelling, we can look forward to the development or the idea for the development of a particular genetic characteristic in livestock and its introduction in large scale into herds of animals.

We can look forward to the introduction of growth-accelerating genes into swine and chickens, steers, and even fish. A growth hormone gene from a trout has been injected into a carp. The carp reached the size of a comparable 18-month-old fish in about 1 year. It is very clear that this technology will have implications for the aquaculture of fish. Such genes, starting with growth hormones, should increase feed conversion efficiencies, improve the quality of the product by decreasing the fat and lean ratios, and allow animals to reach a marketable size much more quickly.

This technology will also affect animal health. Genes that carry immunity to common viral and bacterial diseases will be introduced. Animals will be used to produce medically valuable products. We might, in fact, look forward to the opening of a new industry in which biologically or medically active hormones, biologic reagents, and enzymes will be produced either by plants or by animals. We can, for example, alter the quality of milk protein to produce cheese of various types that might better suit our purposes. We can, as I described, look forward to altering the quality of wool, altering the quality of silk, even altering pearl production in oysters by using this technology.

What about the safety of these techniques? Such changes have been occurring spontaneously in nature for many millennia. This fact is not fully appreciated. The recombinant DNA technology can accomplish its goals much more quickly and with much greater precision than the natural process of gene mutation and genetic assortment that have occurred in nature. Once again, the problem of public acceptance, the problem of public perception is real, and I would expect it will remain one of the principal barriers to the development and investment in this technology. It will probably vary from country to country as each industrialized nation seeks its own opportunities. The United States operates in a global economy in a world in which its own biases and prejudices cannot be easily exported.

We will undoubtedly see a situation in which the development of this technology either occurs in this nation as a major industry or it will occur elsewhere, either with us or without us. This fact, notwithstanding the climate of regulation, the development of a patent policy as it effects these matters, will continue to be important. Ideas are neutral in their value, but not in their implementation. New knowledge will be developed. It will be up to us to use it for the betterment of mankind and mankind is obviously better served by a more efficient and a more available food supply.

BACKGROUND REFERENCES

Sinn E, Muller W, Pattengale PK, Tepler I, Wallace R, Leder P. Co-expression of MMTV/v-Ha-RAS and MMTV/c-MYC genes in transgenic mice: synergistic action of oncogenes in vivo. *Cell* 1987;49:465–75.

Stewart T, Pattengale PK, Leder P. Spontaneous mammary adenocarcinomas in transgenic mice that carry and express MTV/myc fusion genes. *Cell* 1984;38:627–37.

PART III

THE FUTURE OF NUTRITION

ABSTRACT

Professor Kornberg points out that in their excitement about breakthroughs in genetic science, talented young researchers are neglecting the science of nutrition. Yet, we are still ignorant of many of the essential facts about our body chemistry and the effects of nutrients on it.

It is easy to see why nutritional science has been neglected. Compared with the spectacular discoveries of new enzymes or genes, the problems of nutrition are complicated and unglamorous. A controlled study of the influence of a dietary ingredient—be it cholesterol, salt, or sugar—on a disease that develops over decades, is exceedingly difficult because of all the complicating factors like heredity, exercise, infections, and caloric intake. Each of us has different levels of enzymes and proteins, different metabolisms, and therefore different nutritional requirements.

Considering the importance of definitive dietary information for health and economic welfare, Kornberg contends, nutrition cannot be left to the art of medicine or the exploitation of hustlers. The only answer is hard science. It may seem odd to the lay person, but basic research is often the most direct path to solving practical problems. The major discoveries in medicine—x-rays, penicillin, the polio vaccine, recombinant DNA—have all come from pure curiosity about scientific questions, unrelated to specific medical problems.

Kornberg makes a plea for more federal support of research, using the NIH as a model grant-giving institution. He points out we also need support from private enterprise, but that we must beware of the dangers of employing university scientists for commercial objectives. In the rush for biotechnology patents, companies believe it is in their interest to keep research efforts secret.

Kornberg argues urgently that the common industrial wisdom that demands secrecy actually impedes progress and profits no one.

Kornberg concludes by offering a new agenda for nutrition in which diet is tailored to the heritable traits of the individual. In the future, scientists will be able to take the equivalent of a fingerprint for our body chemistry, telling us more precisely the balance of nutrients needed by the individual. And, he emphasizes that the true understanding of nutritional problems, like all biological phenomena, will be through the language of chemistry.

Professor Jules Hirsch of The Rockefeller University says that one priority of scientific research should be to detail the full diversity of human nutritional needs. About 15% of Americans are candidates for high blood pressure, but only some of them are sensitive to dietary salt. To put all Americans on a reduced salt diet will not be necessary. The same is true for cholesterol and fat. Hirsch says that in the 21st century, we will see not one but many sets of dietary guidelines, catering to the special needs of each of the many subgroups that make up our population. In contrast to Kornberg, he emphasizes that understanding human biological problems, such as human nutrition, through chemistry is necessary, but not sufficient. There is also a need for proper clinical studies. In fact, man remains the appropriate test object of any clinical hypothesis about his nutritional needs.

Scientists are understanding more and more about intolerance for certain foods owing to the interaction of man's enzymatic make-up established during evolution and his recent wandering and changes in diet and environment. Intolerance to lactose, the sugar of dairy products, and a certain constituent of wheat, gluten, results in food intolerance. According to Norman Kretchmer, a professor of nutritional science at Berkeley, the food industry is already beginning to respond by designing special foods, like lactose-free dairy products, to satisfy specific nutritional needs.

Irwin Rosenberg, who directs the USDA Human Nutrition Center on Aging at Tufts University, focuses on the special food needs of the elderly. For example, older people use protein less efficiently, so that they risk stressing their kidneys if there is too much protein in their diet. Rosenberg points out that in the early decades of the next century, the proportion of the U.S. population over age 65 will reach 20%, compared with 11% today, making it incumbent upon researchers to understand the relationship between aging and nutrition.

Exploring questions of consumer psychology, Judith Stern, a professor of nutrition at the University of California at Davis and consumer columnist, contends that many of the public's fears about food safety are caused by a failure to grasp the concept of relative risk. In 1989, many of the same consumers who avoided apples exposed to Alar (extremely low health risk), also "forgot" to wear seat belts (moderate risk), smoked cigarettes (high risk), or went hang-gliding (very high risk). Stern urges the scientific community, food manufacturers, and the government to communicate the idea of relative risk to consumers. Stern also warns the food industry that the consumer's demand for convenient packaging is about to collide with the crisis in solid waste disposal. As the United States runs out of landfill sites and the cost to taxpayers of garbage disposal escalates, there will be pressure on the food industry to use less packaging, and to use biodegradable and recyclable materials.

CHAPTER 9

Understanding Life and Nutrition as Chemistry

Arthur Kornberg

N utrition is a keystone for health, and yet as a science, it emerged less than a century ago. When we reflect on the history of medicine in the 20th century, we realize that the *microbe hunters* dominated the end of the 19th century and the first two decades of this century. These hunters tracked down one after another of the microbes responsible for the scourges that had plagued the centuries before it: tuberculosis, cholera, diphtheria.

But there remained some terrible diseases for which no microbe could be incriminated: scurvy, pellagra, rickets, beriberi. Then it was discovered that these diseases were caused by lack of a vitamin, a trace substance in the diet: vitamin C for scurvy, niacin for pellagra, vitamin D for rickets, thiamine for beriberi. These diseases could be prevented or cured by a trace amount of a pure chemical compound and so in the decades of the 1920s and 1930s, nutrition became a science and the *vitamin hunters* replaced the microbe hunters in the spotlight.

In the 1940s and 1950s, the biochemists strove to learn why each of the vitamins was essential for health, and they discovered the key enzymes in metabolism that depend crucially on one or another of the vitamins to perform the chemistry that provides cells with energy for their growth and function. Now these *enzyme hunters* occupied center stage.

Even laymen are aware that the enzyme hunters have had their day. For the past three decades, they have been replaced by a new breed of hunters. The most popular scientific safaris have been those to search for the genes, the blueprints of the enzymes and for the defective genes that cause inheritable diseases, diabetes, cystic fibrosis, Huntington's chorea. These *gene hunters* or genetic engineers use recombinant DNA to identify and to clone a gene. They introduce genes into bacterial cells to create factories for the massive production of precious hormones and other proteins. Genes can also be engineered to endow cells with the capacity to make improved enzymes and proteins.

In view of the inexorable progress of science, we can expect that the gene hunters will be replaced in the spotlight and so we might wonder when and by whom. What kind of hunter will dominate the scene in the last decade of our waning century and in the early decades of the next? Perhaps it will be the neurobiologists who apply the techniques of the enzyme and gene hunters to the functions of the brain. In that case we shall call them the *head hunters*.

I joined this march of medical science in 1942. I had completed an internship in medicine, enlisted in World War II and served in the navy as a ship's doctor. Then I was transferred—I should add with the eager consent of the captain of my ship—to the National Institutes of Health in Bethesda. There I was assigned to do research in the nutrition section. My problem was to determine why rats consuming a minimal diet and given a sulfa drug, developed blood diseases that included anemia, the loss of white cells, and a defect in blood clotting. I learned that sulfa drugs kill off the good bacteria in our intestines that manufacture the vitamins that augment the supplies in our diet. As a consequence, the rats consuming sulfa drugs and the minimal diet became deficient in vitamin K that they needed for blood clotting. They also lacked a vitamin that is essential for forming red and white blood cells and is normally supplied in the diet by liver, vegetables, and yeast. This vitamin was later identified as folic acid.

At the conclusion of World War II, I had spent 3 years feeding rats various diets dosed with sulfa drugs or other agents. I realized that the heyday of vitamin hunting was over, and I was bored. I had entered nutritional science in its twilight. Virtually all the vitamins had been discovered and the new excitement now lay in finding out how they worked and discovering the enzymes that used each of these vitamins as cocatalysts in cellular metabolism. Beyond these new frontiers lay the discovery of endless more enzymes and genes.

Genetic engineering and associated techniques represent the most revolutionary technological advance in the history of biological science. The effects of this advance on medicine, agriculture, and the chemical industry have not been exaggerated. Yet even more profound than the publicized successes of the mass production and modification of hormones and interferons and vaccines is a development that is largely unappreciated. It lacks a name, it has no obvious commercial application. I refer to the confluence of all the basic medical and biological sciences into a single discipline. This coalescence is based largely on their expression in a common language, the language of chemistry.

Let me explain. During the 20th century, in which one and then another discipline of medical science dominated, research and teaching in the sciences basic to medicine were carried out in an increasing number of discrete departments. By the time I entered medical school, some 50-plus years ago, there were departments of anatomy, biochemistry, physiology, bacteriology, and pharmacology. They were as separate as the departments of chemistry and physics and biology were then and still are today. Departments of genetics and neurobiology did not yet exist in medical schools.

The situation is radically different today. Research and teaching in all these departments are interdependent and virtually indistinguishable. Consider anatomy, the most descriptive of these sciences and then genetics, the most abstract. They have simply become chemistry. Anatomy can now be studied (and should be studied) as a continuous progression from molecules to the tissues that make up a functioning organism.

Transformation of genetics has been far greater. Whether heredity operated by known physical principles was a serious question only 40 years ago. Of course, we now understand and examine genetics and heredity in simple chemical terms. You have heard and know that chromosomes and genes are easily analyzed, synthesized, and rearranged. We can modify a species at will. Thus, whether we can sequence the 3 billion units that make up the human genome is no longer a question, but rather, what government agency will sponsor it and what it will cost. Nevertheless, the importance of chemistry, as the foundation of all medical science, is usually submerged and obscured by attention to specific and urgent problems. This simple truth may escape both physicians and scientists.

Physicians are inclined to action, especially surgeons, and there is a story often retold of the surgeon, who, while jogging around a lake, spotted a man drowning. He pulled off his clothes, jumped in, dragged the victim ashore and resuscitated him and then wearily resumed his jogging, only to see another man drowning. So, he repeated the performance and by this time he was pretty exhausted only to look up and see several more people in the lake drowning. He also saw the professor of biochemistry, who was absorbed in thought. He called to the biochemist to go after one, while he went after another. When the biochemist was slow to respond, he shouted, "Why aren't you doing something?" The biochemist said, "I am doing something. I am desperately trying to figure out who is throwing all these people in the lake."

This parable is not intended to convey a lack of regard for fundamental issues by physicians nor callousness by biochemists. Rather, it portrays the reality that the war on disease must be fought on several fronts. Some must contribute their special skills to the distressed individual, while others must try to gain some broad knowledge, a knowledge base necessary to outwit both present and future enemies.

Even scientists, as I have said, fail to appreciate the importance of chemistry. Physicists are interested in the behavior of nuclear particles and cosmic phenomena. They are generally bored by molecules and by chemistry. Biologists know that molecules determine the shape and the function and the fate of cells and organisms, but they shy away from the complexity of body chemistry. As for the general public, distinctions among the natural sciences have very little meaning. Many people cannot distinguish an atom from a molecule, a virus from a cell or a gene from a chromosome. So, it is no wonder that journalists and lawmakers keep asking us whether a molecule, a virus, or a cell is living. They become impatient when we fail to give them a straight and simple answer as to when and where the breath of life enters or leaves.

Unfortunately, fashion prevails in science, probably as much as in other human activities and the tides of fashion erode one beach to create another. So, with the appearance of a golden age of new discoveries, there is inevitably a neglect of the skills and insights of the age that is eclipsed. So, valid approaches are ignored and important questions are left unanswered.

An eminent Swiss philosopher physicist, Marcus Fierz said, "The scientific insights of an age shed such glaring light on an area as to leave the rest in even greater darkness." In the rush and excitement over the discoveries of enzymes and genes, the science of nutrition was abandoned and major problems in human nutrition were grossly neglected. In the last 50 years, relatively little has been added to the core of nutritional knowledge. Despite the heroic efforts of a few individuals, including some at this conference, the science of nutrition is mostly in shambles. Ignorance and fear of chemistry distort our lives, particularly in our nutrition. We are the victims of ignorance of essential facts about our body chemistry.

One of my pet peeves for years has been the unsubstantiated dietary advice with which we are bombarded, about cholesterol, for example. Signs in the Stanford University hospital cafeteria admonish us, "Don't eat eggs. Don't eat fat. Don't eat sugar. Don't eat salt. Eat and you die."

I was delighted to see the lead article in an issue of the *Atlantic Monthly*, which is an excerpt of a book by Thomas Moore, in which he cites a "cholesterol myth" that needs debunking. Lowering cholesterol by diet is unlikely and cholesterol-lowering drugs may even be dangerous.

Proof that dietary levels of cholesterol have a significant influence on the blood levels of cholesterol and the incidence of heart disease in the majority of the population has not been established unequivocally. Having read the Moore article and some of the original literature, I believe that except in cases of a rare genetic disease and a limited fraction of the population, available data do not substantiate claims that the reduction of dietary cholesterol significantly reduces tissue and blood cholesterol levels, despite the assertions of prestigious and well-meaning organizations, such as the American Heart

Association and the National Heart and Lung Institute. Also, inadequate information is available regarding the danger of reducing cholesterol, an absolutely crucial body constituent.

Of course, people want to know what and how much to eat. Yet, few people realize how little is known about nutrition. Most people have no way of separating available knowledge from the faith healing and quackery that masquerade as science. Turning to physicians is of little help because they don't know much about nutrition either, and one cannot teach what one does not know. It is no wonder that people will grasp at how to achieve health and longevity by adherence to a few simple dietary commandments.

What can be done to restore nutrition as a science and to advance nutritional knowledge? My plea is that much of life and certainly nutrition can be understood in rational terms if expressed in the language of chemistry. It is an international language. It is a language for all of time. There are no dialects. It is a language that explains where we came from, what we are, and where the physical world will allow us to go. Chemical language has great esthetic beauty, and it links the physical sciences to the biological sciences, our past to our future.

The recognition of these simple truths unfortunately is not common. You recall the slogan of the Dupont Company for many years when it advertised, "Better things for better living—through chemistry." The purpose of the slogan's message was to inform the public of the value of plastics, herbicides, and industrial chemicals for our individual and collective well being. The campaign, I assume, was successful, for a time at least, in promoting good will for Dupont and for the chemical industry. Today the slogan is simply, "Better things for better living." "Through chemistry" was dropped when it became known that these chemicals, as is true of all things, whether they are natural or manmade, can be toxic. Some time ago in the food market, I overheard a little girl saying to her mother, "Mommy, you should not buy that. It has chemicals." I was startled but the mother was not, nor was the store clerk. They did not find that remark strange or disturbing.

There has been no advertising or educational campaign, no excellent television program, that has convinced the public that life is a chemical process. The public has not been taught that the human organism, its form and behavior, are determined by discrete chemical reactions. The origin of the human organism, its interactions with the environment, and in important respects, its fate is dependent on chemical processes. In fact, lately the only time I have heard chemistry referred to as good is when people speak of the success of some collective social effort, such as a winning football team or a business group having the "right chemistry."

The waning of nutritional science and its depopulation are understandable. Compared with the spectacular discoveries of the vitamins that prevented and cured classical deficiency diseases, the residual problems of animal and human nutrition are elusive. Pursuing them scientifically is difficult, time consuming, and unglamorous. Consider, for example, the consumption of sucrose, which increased many fold in western societies at the same time a comparable increase occurred in the incidence of heart disease. Were they causally related? A controlled prospective rather than retrospective study of the influence of a dietary ingredient, be it sucrose, cholesterol, or salt, on a disease that develops over decades, is exceedingly difficult especially when many factors such as heredity, exercise, infections, and caloric intake, may all contribute in significant ways.

Similar difficulties arise for those who try to determine the undoubtedly important effects of diet in the prevention and treatment of cancer and heart disease and immune disorders and the influence on aging. These are exceedingly complicated and perplexing processes and they create a difficult setting for the growth of a struggling scientific discipline.

A major factor in nutrition, aside from all of these environmental influences and the intrusion of diseases, is the individuality of our organ systems. The genetic distinctiveness that is obvious in our faces and our fingerprints is, of course, dictated by our DNA blueprints and it is well known to anatomists and surgeons that the shape of our body organs is also unique. When one examines 12 human livers in turn, not only do they have different shapes and sizes, but they range in weight by a factor of 4. Twelve real stomachs examined in turn all differ from each other and the "textbook" stomach; these stomachs do not look, nor would they be expected to work, alike.

What is seen in this variety of shapes of livers and stomachs is a reflection of the slight but significant variations in the design and abundance of enzymes and proteins in the cells of these organs. It follows without question that each one of us has unique features in our nutrition. The individuality of the digestion and metabolism of each of us profoundly affects our choices of foods and our responses to them.

Advances in technology have also complicated human nutrition by giving rise to opportunities and questions that were previously beyond reach. One such advance is the abundance and cheap supply of pure vitamins that has made megavitamin dosage a feasible and common practice. Does a vitamin intake 100 times that needed to maintain normal levels provide some benefit, even to a small percentage of the population? Are there circumstances in which such megavitamin dosage may also have adverse effects? I find it hard to believe that huge amounts of a chemically reactive substance such as vitamins A, C, and D that may help a few individuals will not harm a like number.

Other scientific advances have given us exceedingly sensitive tests that determine whether food may cause mutations and cancer. We now learn that a large number of natural foods and food processes can be mutagenic, and they can be presumed, therefore, to be carcinogenic and perhaps to accelerate aging. This long list includes mushrooms, celery, chocolate, bruised potatoes, alfalfa sprouts, black pepper, peanut butter, herb teas, charred meat, and a hundred other common foods, not to mention additives, preservatives, and coloring dyes.

To monitor an expanding number of incriminated foods, to assess properly and soberly the significance of these so-called mutagenic levels, and finally to formulate a varied, palatable, balanced diet for different age groups in assorted states of health and disease, to do all this, exceeds the limits of available resources and knowledge.

Perhaps the most perplexing problem we face in human nutrition is the confusion between nutrition as a natural science and nutrition as the social, political, and economic science of feeding people. It is a confusion analogous to that between medical science and health care. Widespread hunger that prevails in developing nations is intolerable. Thousands of children go blind in India each year from a deficiency of vitamin A that could be prevented by a few cents' worth of that synthetic vitamin. All 10 essential vitamins and 8 minerals can be supplied at a bulk cost of a dollar or so per person per year.

Widespread hunger and deficiency diseases are failures for which society, not the nutritional scientist, is to blame. In overreaching its scientific domain, nutrition like so many other human endeavors becomes prey to prejudice, chicanery, and other perils that beset a culture. Nutrition, when it is compelled to encompass medicine and agriculture, economics and psychology, anthropology and dietetics, as well as chemistry and biology, becomes more a social and political activity and less a science.

Whither human nutrition? How can we cope with the numerous and complex problems and the extraordinary difficulty of doing controlled, long-term dietary experiments on humans? Considering the importance of acquiring definitive dietary information for health and economic welfare, we must agree that nutrition cannot be left to the art of medicine or the exploitation of hustlers. The only answer is science. Hard science. Progress demands these actions.

As I have suggested in the past, we must (1) invest in the training and support of scientists to work in nutrition, (2) narrow the focus of experimental work to small, doable problems whether or not they seem urgent, (3) use a variety of animal models, (4) exploit the advances of biotechnology to advance the chemistry of metabolism, and devise quick and sensitive assays of an individual's metabolic processes, (5) insulate nutrition from social issues, and, finally, (6) sustain the faith that persistent, creative scientific effort eventually solves most problems, often in a surprisingly early and novel way.

I would pointedly avoid a crusade on this or that nutritional problem. Crusades on disease fail. They fail because they lack the weapons of basic understanding. In urging a course of action to solve a major medical problem, it often seems counterintuitive both to scientists and to lay people that the most direct path is basic research.

Investigations that may seem irrelevant to any practical objective are almost invariably the path to major discoveries in medicine, for example, x-rays, penicillin, polio vaccine, recombinant DNA. All these and more have come from the pursuit of curiosity about questions in physics, chemistry, and biology apparently unrelated at their outset to any specific medical, agricultural, or industrial problem.

What is truly unique about science is not the people. It is the discipline that enables ordinary people to go about doing ordinary things that, when assembled, reveal the extraordinary intricacies and the awesome beauties of nature. Science not only permits these people to contribute to such grand enterprises, it also offers them a changing and endless frontier for exploration.

For the pure and applied sciences to develop in academic centers, it is crucial that they have the generous support of the federal government and that they develop strong ties with industrial enterprises. With regard to federal support of research, we can look with great pride and confidence on the achievements of the National Institutes of Health. The NIH expends nearly 80% of its budget in grants to individuals and private research institutes. The awards are based on peer review by panels drawn from outside the government.

Major credit for the extraordinary revolution in biologic science belongs to the NIH. As expressed by Louis Thomas, "all by itself, this magnificent institution stands as the most brilliant social invention of the 20th century anywhere.... It has been unique, imaginative, useful, and all together right."

Were the NIH record to be put forth as an experiment in research administration, an impartial observer in this social area of science might well question whether other factors might have been responsible for success. An experimental control exists in the support programs of agricultural science in the United States during the same postwar period. The Department of Agriculture, by contrast with the NIH, retained all authority within its own bureaucracy and limited research activity to established regional laboratories around the country. There were few grants to universities and private institutions.

With this old-fashioned system of research management, the knowledge base for agriculture remained relatively static. Too little was learned about the basic biochemistry and genetics of plants and farm animals. In fact, some of the basic plant research during this period was bootlegged with NIH grants. Only recently, with the introduction of recombinant DNA technology, has there finally been some awakening of interest and activity by the Department of Agriculture in supporting basic agricultural science.

With regard to academic–industrial relations, I am aware of uneasiness about biotechnology. One must recall that the scientists who provided the ideas and the reagents, the techniques and machines, and the very practitioners of genetic chemistry and immunology came exclusively from universities. We scientists and our universities are reluctant to be excluded from the financial rewards that are very much needed to continue research, and there are obvious dangers if a university, as a nonprofit corporation, becomes entrepreneurial and then employs its faculty and students for commercial objectives. There are major dangers, too, if biotechnology and food and drug companies were to appropriate a generation of senior scientists, use them as executives and consultants, employ junior scientists, and then seal them off from a free exchange of new knowledge.

There can be no question about the desirability and urgency of putting this new knowledge to practical use, nor is there any question that strong connections between academia and industry are needed to ensure the growth of the drug and food industries. Intimate relationships over many years between university chemistry departments and chemical companies created a vast industry in this country; the same is true for physics and engineering departments in relation to creation of the electronics industry.

The unusual feature of industrial applications of biological science is that the basic technology came exclusively from academic laboratories and was developed rapidly. This is the reason that university scientists and administrators are concerned and often vocal. Related to this concern about academic–industrial relations in these industrial developments is the danger of secrecy. Most scientists are convinced that secrecy in an academic setting impedes progress and doesn't profit anybody. University and business leaders have agreed on occasions when they have met to consider this issue that corporate-sponsored research in academic centers should be open. Well, sort of open. I would go much further. I believe that secrecy makes even less sense in an industrial setting than in an academic setting. If you are surprised at this, I would remind you that an academic scientist can justifiably fear that the release of an idea, a hint that something works, the source of a key reagent, would enable an academic competitor to publish quickly and gain priority for an important discovery.

Appropriation of a new idea is far less likely in an industrial enterprise. Developing a major product, especially a pharmaceutical or food product, requires approval of the FDA. It requires commitment to spend a lot of money over many, many years and for this reason, many factors other than having the original idea, matter more. The shrewd choice of a goal, a high competence in attaining it, and finally, and perhaps most important, an effective marketing of the product.

The best insurance for success in industry is an open atmosphere, one that provides optimal access to all available information and advice. As significant advances are made, the inventions and discoveries can be protected each step along the way by a patent application team in which the lawyer interacts on the spot continually and knowledgeably with the scientist.

By contrast, a company policy of secrecy that closes the avenues of exchange with academic scientists prevents useful collaborations. It is generally unappreciated that secrecy within the company shelters mediocrity and discourages the vigorous exchange that identifies excellence.

Scientific truths are logical. They can be stated simply and they are hard to conceal. By contrast, technological developments are more cultural and more intricate; in fact, it may be hard to give them away. For these reasons it is possible for Japan, with relatively few contributions to basic science, to readily assimilate that knowledge and then excel in its application. England, by contrast, despite preeminence in basic science, has lagged behind in technological developments.

The argument against industrial secrecy becomes even more compelling when it relates to a field that has enormous potential for growth. The impact on medicine and industry of biotechnology, will, I believe, exceed our most optimistic claims. Of course, there will be fluctuations in enthusiasm, particularly of venture capitalists, reflecting their appraisals of how soon and by whom money will be made.

My bullishness for the power of this new technology to make practical advances in medicine and industry remains strong because I am impressed by the impact of genetic engineering and the new immunology on basic biology. From our deeper understanding of the organization and functional control of chromosomes will come opportunities for intervention in aspects of growth, development, and aging that are now impossible. Within a few years, the most exciting prospects for medicine and industry will be subjects and products that no one has yet thought of.

One may wonder which scientists will make these new discoveries and what organizations will be equipped to give them opportunities to be creative. My experience is that the scientists who will make these new advances are already bored with available procedures and are not eager to join the pack to clone another hormone or make a better vaccine. Instead, they seek new techniques to solve more challenging problems.

As in the past, I believe these pathfinders in biology will work in academic institutions and may be completely insulated from industry. Yet, this pattern can be changed. I believe that wise and farsighted management of an industrial enterprise can help effect this change. Able scientists have lately become far more interested in industry. They are becoming discouraged by the atmosphere in university departments, where research support has become severely straitened, where there is an emphasis on fashionable areas (e.g., AIDS), and where entrepreneurial grantsmanship is required. They are unhappy with academic bureaucracy, service on committees, heavy teaching loads, and the pressure to choose some fashionable research program that will produce publications for academic promotion and the next grant application.

In the face of these problems, one might find in an industrial setting several clear advantages: excellent resources, research objectives in an interesting area of science, fewer distractions, and often a team spirit for achievement in which all can share success. The critical ingredient that must be provided by industrial management, if it wishes to capture and retain creative and productive scientists, is an open atmosphere, one which encourages the scientists to discuss ideas, progress and failures with colleagues, in and out of the organization and to publish without restraint. Such a company atmosphere is conducive to a flow of students, postdoctoral fellows, and visiting professors, through its laboratories. It is an atmosphere in which the scientist can feel that his or her creative ability is being fostered. I know that this system has worked when given a proper chance.

To return to food and nutrition in particular, I have touched on a variety of issues that make progress in this area a most perplexing problem for science and society. Faced with the bewildering enormity of the problem, I am reminded of the poor rabbinical student who appealed to the lord. "Lord," he said, "I simply can't grasp the age of the universe. The scientists tell us it is hundreds of millions of years old." And a voice came down, "Relax, my child. In the order of things, millions of years are only a few moments." "Thank you, Lord. You know, I have the same problem imagining millions of dollars." Again, the voice, "Simple, my child. Those sums are also trivial, like a few pennies." To which the student brightly responded, "Lord, may I have one or two of those pennies." "Why, of course, my child," the Lord answered, "just wait a few moments."

BACKGROUND REFERENCES

Kornberg A. *For the Love of Enzymes: the Odyssey of a Biochemist*. Cambridge MA: Harvard University Press, 1989.

CHAPTER 10

A Brief History of Nutrition: The Role of Clinical Science

Jules Hirsch

I remember a tale told by a distinguished neurophysiologist, who was trying to summarize the nature of our thought and culture and formulate a way to present it to some extraterrestrial beings. His idea was to prepare an extraordinarily powerful spaceship, load it with the contents of the Library of Congress, and shoot it out to the beings of extragalactic space. They would figure out just what we are up to and make contact with us.

Ultimately, the space ship did arrive in a place with creatures using brains made out of silicon perhaps with little bits of germanium, all working in some way without using or knowing of books. But they set to work, looking at the books and trying to figure out their nature.

The first thing that they did was to call their chemists. Chemists were very important in that society, and what they did first was burn one of the books to determine its content. The ashing of course gave an elemental analysis. Then after years and years of approaching the problem through chemistry, their local equivalent of our Nobel laureates won their special prize for uncovering the structure of cellulose, the basic constituent of paper.

They had no understanding of the purpose of the books, but they learned a good deal about the structure of cellulose. I offer this parable as an example of totally missing the point through the use of chemistry, rather than creating better things by the use of chemistry. It is possible to miss central issues by using chemistry, when chemistry alone is not an appropriate framework for understanding.

Chemistry must be applied to study nutrition, but from the standpoint of the physician or the clinical investigator chemistry must be used selectively.

Man maintains a fairly fixed composition of body fat and body minerals throughout their lifetimes, regardless of materials that pass through the body. The exchange of body substance with the world reminds us of our close link with the world. The goddess, Ceres, who has been chosen as the symbol of this meeting, is a symbol of man's indissoluble link to plants and fields.

There is a lovely statue of Ceres in the Abby Aldrich Rockefeller garden here in Williamsburg. She is a beautiful woman, also part plant. There are sheaves of wheat in her hair and in her arms. Our relationship with nature through her was initially through propitiary sacrifices and then by the creation of our beautiful gardens and vineyards. The special meaningfulness of the farm in our culture remains as a tribute to Ceres and our link to "mother" nature. Dubos referred to this special relationship as man's wooing of the earth. Although our relationship with Ceres has changed through scientific insights beginning in the 18th century, she will still be with us in the years to come. Her presence will be manifested in our special relationship to plants and flowers. No scientific principles are apt to abolish our deep relationship with Ceres that has endured since antiquity.

Radically new thoughts on food and nutrition came in the late 18th century when the scientists and philosophers of the enlightenment examined the nature of combustion. In one of the greatest experiments, Lavoisier placed a guinea pig in a chamber and showed that it was giving off heat as measured by the melting of ice. The amount of carbon dioxide that it liberated or oxygen that it took up had an exact proportionality to the rate of heat production, as measured by the water dripping into the receptacle. It was then shown that combustion is precisely the same whether wood is being burned or combustion occurs in a guinea pig. Combustion is the same process in a manmade fire or in living tissues. There is no vital heat. This observation was a great blow to vitalism and it marked the beginning of the study of foods and animals in a scientific and predictable way. Thus, our intimate relationship with Ceres began to be colored by scientific observation.

For 100 years or more, much of the science of nutrition, especially in animal husbandry, was based on these principles of calorimetry. Initially, it was believed that it made no difference what was burned, that is, that what one eats made no difference. Then came the understanding that specific vitamins and other micronutrients were uniquely necessary for the continued health of this organism. Furthermore, certain kinds of protein were specifically needed, which was an additional discovery, equally as important in some ways as the discovery of the vitamins.

Their findings colored the history of life sciences in the early 20th century. In fact, these discoveries marked the beginning of modern biochemistry and formed the scientific underpinnings of the National Institutes of Health. These are the discoveries that led to much of the original development of NIH.

In the latter half of this century there was a departure from looking for specific nutrients to a broader search for diets most consonant with long-term health. In many respects, this concern evolved because populations were beginning to age. In this part of the world, people were increasingly concerned with degenerative diseases, the expense of medical care, and its relative scarcity, and thus, the preventive medicine movement came upon us. This intersected with early and incomplete nutritional information, and we became launched on hucksterism, hope, and approaches beyond the scientific pale.

Modern nutrition has been trying desperately to find the best diet for us. I am fearful that further clinical investigation in nutrition will be foreclosed, however, by the belief that all answers are available through chemistry alone, and there is no more clinical research to do. Today, studies are focused on intricacies of the nucleotide sequence. I believe there will be more and more to be learned from intact man, as we unravel our nucleotide sequences.

Much of modern biomedicine came about through the study of man in health and disease. Observations on human pneumonias and the virulence of the pneumococcus led to the great biological experiment of the 20th century, the discovery of the transforming principle, DNA as the genetic substance. Curiosity and questions about man and animals led to the great 19th century experiment of Lavoisier that opened up the possibility of modern biochemical research. I believe that further study of the problems of human nutrition in intact man can generate similarly consequential observations.

What is the future agenda for human nutrition? To begin, we must understand the diversity of man in terms of nutritional needs. The relationship between salt intake and blood pressure is a case in point. Fifteen to 18% of Americans develop hypertension. This subset of the population is salt-sensitive and would benefit from less salt in their diet to lessen their blood pressure. It is difficult to detect hypertension, because people may not go regularly to their physicians for blood pressure measurement. So it would be ideal for those with salt sensitivity to begin early treatment by reducing salt intake. Nevertheless, imposing a low-salt diet on the other 85% of Americans is unnecessary. If we had the means to determine who might and who might not become hypertensive, we could treat those at risk at an early stage. Describing such differences in human nutritional needs is an agenda for the future.

We are sufficiently different from each other in habits, metabolism, and energy utilization to require different levels of caloric input and different kinds of diets. The clinical investigator will have to use the best available methods of modern biochemistry and biotechnology, but also chemical observation and sound judgment to specify our needs.

Another concept for consideration is an "anthropic principle." The nutritional anthropic principle is as follows: our dietary selections, the foods for which we hunger, relate to our very distant evolutionary past and are now our uniquely human heritage. For example, humans are well built to prevent starvation; a fraction of 1% of everything we consume is sequestered in a knapsack called adipose tissue. This occurs in preparation for starvation. Were no food available, we could survive for 30 or more days. If we are obese, we would survive longer. Some of us have lineages that suffered with famine and, as a consequence, are better padded with fat than others. With the abundance of food of different types and the relative absence of starvation in our western world, this special genetic endowment has become a disease. The anthropic principle drives our food choices. We are not pleased unless we eat sweet rather than bitter foods since this usually means more rather than fewer calories. This is our nature and is not apt to change in the near future.

The food industry and the clinical investigators of nutrition have an obligation to examine these problems and to develop a variety of foods through which the diversity of our genetically programmed needs can be met. This could give us the satiety and "quality of life" coming through our anthropic relation to foods, yet make it possible for us to lead healthy lives.

We are now at the beginning of nutritional science that can draw on the tools of biotechnology. But the central figure of nutritional science must remain the investigator who wishes to study human beings and their problems in all dimensions: chemistry, our myths about "Ceres," our evolutionary history, and our present behaviors and needs.

BACKGROUND REFERENCES

Burton BT, Foster WR, Hirsch J, Van Itallie TB. Report of conference proceedings on the health implication of obesity: an NIH consensus development conference. *Int J Obesity* 1985;9:155–69.

Hirsch J, Fried SK, Edens NE, Leibel RL. The fat cell. In: Bray G, ed, *Obesity: Basic Mechanisms and Clinical Applications*. Philadelphia: Saunders, 1989.

Hirsch J, Leibel RL. What constitutes a sufficient psychobiologic explanation for obesity? In: Stunkard AF, Stellar E, eds, *Eating and its Disorders*. New York: Raven Press, 1984: 121–130.

Leibel RL, Hirsch J. Metabolic characterization of obesity. *Ann Int Med* 1985;103: 1000–2.

CHAPTER 11

The Evolution of Diet and Nutrition

Norman Kretchmer

I would like to present two stories, in a sense two anecdotes, one in which the science is almost completed, and the other in which, although there has been a great deal of work, we are just beginning to understand the problem.

Food is an extreme environmental pressure and has influenced our development over the millennia. Modern man is believed to have had his beginnings between 50,000 and 100,000 years ago. His environment then was vastly different from ours, and, consequently, his diet was not comparable. Early man probably originated as a gatherer eating mushrooms and roots. As courage increased and tools were improved, some emerged as hunters. A few of these prehistoric societies still exist in the world: outstanding examples are the !Kung of South Africa, some isolated Eskimos in the far north, some tribes in the Philippines, and the Pygmies.

It took approximately 90,000 years of human cultural development for agriculture to emerge. It did so in a number of discrete locations throughout the world. In certain areas, such as the Andes and Indonesia, new edible plants were developed. In others, such as the Euphrates River valley, there was careful domestication not only of plants, but also of animals. As new foods entered the diet, metabolic adaptations occurred in man. During the Neolithic period, man divided into two general agricultural groups: farmers, who domesticated, developed, and grew crops were really the first plant breeders and pastoralists, and those who were animal husbandrymen, who raised cattle and other animals and began wandering, looking for pastures for their herds. The diets of these two groups were considerably different; the farmer used grains and legumes as his staples whereas the pastoralists used animal products and dairy products as major dietary components.

As millennia have passed and we have come to the modern day, more processed foods and condiments have entered our diet. The result is our current industrialized, processed, and, in some cases, synthetic melange of foods. Think for a moment of the changes that had to take place in our metabolic processes so that we might adapt to these dramatic dietary changes.

Evolution is a complex process, involving an unpredictable, successful, hereditary change in a living organism. There are three main factors involved. The first is mutation, an actual viable genetic change in an individual, who then reproduces that change. The second is selection, a complicated process that involves interaction of the environment with the genetic change, resulting in the proliferation of the fittest. And, finally, the continual outbreeding of these individuals, harboring the successful mutation.

In nutrition, there are many examples of evolutionary adaptation. Two are prominent as particularly clear reflections of these events. One first appeared approximately 10,000 years ago on the Arabian Peninsula, probably in the Euphrates River valley, when man first domesticated cattle and milk became an important food for adults. For unknown reasons, man wandered and migrated to new pastures. Migrations occurred in three major directions: across the then fertile Sahara, into the Sudan and beyond; in the north toward the Northeast, Northern India, and Mongolia; and in the Northwest to Northern Europe. With the cattle and with the men went changes in language.

One of the main foods on these treks was milk and milk products. Thus the ability to digest various constituents of milk as a source of energy and a source of nitrogen was life saving. Almost all milks contain the sugar lactose, which is digested by an enzyme called lactase located in the small intestine.

If the lactase is not present, the lactose is not digested and that sugar then passes intact into the lower bowel, where the bacteria ferment the sugar, causing in its most severe form, a severe fermentative diarrhea. Who can digest lactose? All children up to the age of about 5–7 years and only certain adults, that is, Northern Europeans, Mongols, certain pastoral African tribes and Punjabis of northwestern India.

Most of the peoples of the world do not have the mutation required to produce the enzyme in adult life and, thus, have difficulty with the digestion of lactose.

This mutation for the persistence of lactase in adult life is assumed to have appeared in individuals, probably on the Arabian peninsula 9,000–10,000 years ago. The mutation is estimated to have appeared in about 1 of 100,000 people, as do most mutations and was neutral; it caused no harm whether you had it or not. The mutation nutritionally favored the diet of the pastoralists and the adult milk drinkers. They could drink milk at all ages. This mutation has persisted for 10,000 years. Today more and more mixed groups have an increased ability to digest lactose, for example, the Semites, the American Black, the Mexican–American, and outbred Native Americans.

The fact that many people cannot digest lactose has led to a number of individual events in the dairy industry in the United States and other countries. For example, industrial methods have been developed to remove lactose from dairy products, so that they might appeal to a greater population.

Another example of evolution in food is the appearance of adult diabetes in a number of indigenous groups. Most of us in nutrition believe that diabetes in most Native Americans, Polynesians, and Australian Aborigines is reaching epidemic proportions. Diabetes is also rapidly increasing among the Black and Hispanic populations in our country.

Incidentally, the lowest rates of diabetes in the world are found among Eskimos and natives of the highlands of Papua, New Guinea. In the early 1900s, the U.S. Public Health Service did not report one case of diabetes in the Indians of the tribes near Phoenix, the Pima–Papagoes. Estimates today indicate that in 50 years all of these Indians will have diabetes by the age of 35. The northwest Australian aborigines were free of diabetes 10 years ago, and now of the aborigines living in northwestern Australia, approximately 18% of the population has diabetes. No doubt all of the terrible statistics result from a change in lifestyle. But what precisely does that mean? There have been many changes, from active to sedentary, no alcohol to alcohol, no sweets to sweets, leanness to obesity, the great outdoors to houses. Studies have clearly shown that as more Caucasians breed into these populations, there is a decrease in diabetes. The genetics of the Native American and the Australian aborigine are absolutely unrelated; they derive from different areas and have distinctly different racial backgrounds.

The theory of the thrifty gene promulgated about three decades ago by James Neal from the University of Michigan is widely used to explain the appearance of diabetes in these populations. In short, Dr. Neal proposed that these hunter–gatherers existed on the edge of survival. They had a life of feast or famine with more famine than feast. They had to evolve a metabolism that would store food during the feast, so that energy would be available during the famine. The major animal storage product is fat. The major hormone involved in metabolic storage is insulin. These people developed a hyperinsulinemic diabetes.

Thus, the development of individuals with an elevated, responsive insulin would be the most efficient way to combat famine. The system was superb as long as people were active and avoided obesity. Once the people became sedentary and obese, the elevated insulin was less effective, and diabetes, the inability to utilize sugar, developed with all its subsequent tragedies of neuropathy, peripheral vascular disease, and blindness. The extreme environmental change probably resulted from a change in food supply and lifestyle.

The diet of man has changed dramatically since paleolithic times. The strict hunter ate a diet that was very high in protein and fat and negligible in carbohydrate. The Eskimo is a good example, the hunter Eskimo who ate fat and protein and practically no carbohydrate. The strict gatherer, on the other hand, ingested food very high in carbohydrate and fiber and low in protein and fat. At present, we eat a diet composed of a good deal of simple carbohydrate and fat.

Each of these diets demands different metabolic adjustments. Differences in man's dietary needs will become more apparent as we learn more. We know that there are individuals who cannot tolerate wheat, who have an inability to excrete salt, or who have problems with dietary fat or cholesterol. I believe that 21st century nutritional science will bring more examples to light, more scientific discoveries and explanations, and industry will respond by designing special foods for particular purposes. Such response is apparent today with manufactured foods low in lactose, high in fiber, low in fat, and low in cholesterol.

Individuality, determined by the interactions of the environment with genetics and evolution, to my mind, is the keystone of modern nutritional science.

BACKGROUND REFERENCES

Eaton BS, Konner M. Paleolithic nutrition: a consideration of its nature and current implications. *N Engl J Med* 1985; 312:283–9.

Haffner SM, Stern MP, Hazuda HP et al. Increased insulin concentrations in nondiabetic offspring of diabetic parents. *N Engl J Med* 1988;319:1297–1301.

Howard M, Bennett P. Diabetes and atherosclerosis in the Pima Indians. *Mt Sinai J Med* 1982;49:169–75.

Knowler WC, Pettitt DJ, Bennett PH et al. Diabetes mellitus in the Pima Indians: genetic and evolutionary considerations, *Am J Phys Anthropol* 1983;62:107–14.

Kretchmer N. Food: a selective agent in evolution. In: Walcher D Kretchmer N, eds, *Food, Nutrition, and Evolution*. New York: Masson Publishing USA, Inc., 1981: pp. 37–49.

Neel JV. Diabetes mellitus: a "thrifty" genotype rendered detrimental by "progress." *Am J Hum Genet* 1962;14:353–63.

O'Dea K, Trainides K, Hopper JL, Larkins RG. Impaired glucose tolerance, hyperinsulinaemia, and hypertriglyceridemia in Australian aborigines from the desert, *Diabetes Care*, 1988;11:23–9.

Renfrew C. The origins of indo-european languages, *Sci Am* 1989;261:106–14.

Zimmet P, Dowse G, Laporte R, Finch C., Moy C. Epidemiology—its contribution to understanding of the etiology, pathogenesis, and prevention of diabetes mellitus, In: Creutzfeld W, Lefebvre P, eds, *Diabetes Mellitus: Pathophysiology and Treatment*, Berlin: Springer 1989: pp. 5–26.

CHAPTER 12

Nutrition and the Elderly

Irwin Rosenberg

The interaction between nutrition and aging is complicated and poses a major challenge to those in science, in industry, and to government, particularly in respect to the kinds of public policy, questions which must be addressed.

In the 21st Century, 20% of the population will be over the age of 65. Within the first quarter of the 21st century, 25% of the population in this country will be over age 65, a percentage that has already been achieved in Japan.

This is a striking change from the turn of the last century when 1 in 25 members of our population was over age 65, and life expectancy was 47 years as compared today to over 75. These demographic changes are typical of other industrial nations. Although developing countries still have a much younger population, these same trends toward an aging population have been identified. This massive demographic shift in the course of one century from a population that is mostly young to one that includes a substantial number of older and very old individuals, presents a major scientific and policy challenge.

Not only are there striking differences among people who are 65, 75, 85, and 95, and it is not at all uncommon today for persons to live vigorously into those years, but we also know that a 70 year old may be very different from the 70 year old next door. There is enormous heterogeneity in this population based on both genetic determinism and environmental and habitual behavior. Understanding the factors involved and understanding the extent to which we can influence their interaction is a major challenge. Simply stated, our society and environment must contribute to the numbers of our elders who will lead vigorous, fulfilling, and independent lives.

Nutrition must play an important part in that strategy, just as successful aging has been influenced by the quality and availability of food over the course of this century. We know that if one studies the functions of physiological systems over the decades of life in a cross-sectional manner, a decline occurs in many of them. What we do not know is the inevitability of that decline and the extent to which it occurs in individuals as opposed to populations of 50- or 60-year-old people compared to one another. That certainly will be one of our research challenges.

The change from a young individual whose body composition has a great deal more lean muscle mass to the older individual who has much less muscle and much more fat is characteristic of the problem of aging and the process of aging. The before- and-after effects of putting a young individual in a leg cast would result in some of the same effects. In the process of disuse or immobilization, a similar, although perhaps less dramatic, shift occurs from lean muscle mass to atrophy and replacement by fat and adipose tissue.

As lean muscle mass is lost, the same kind of atrophy occurs in respect to the mineralization of the skeleton. If one casts or immobilizes a leg, skeletal and calcific mass are lost. During the process of aging, a kind of loss takes place.

Aging has been described in the context of a disuse syndrome. Most organ tissues, including the skeletal mass, can be maintained by physical activity and by energy flux through the system. With disuse, muscle mass and skeletal mass are lost, fragility increases, and obesity and cardiovascular vulnerability develop.

The decreased level of activity seen throughout the aging process, which tends to parallel the loss of lean body mass, is not a necessary change in behavior. Although one does not have to exercise vigorously and be hooked up to instruments that measure oxygen capacity, it is important to maintain both fitness and diet to influence the usual aging process.

We must understand more about all these processes, these physiological changes that occur, and determine what to recommend that will influence these processes to enable the population to maintain a more vigorous, active, and independent existence throughout life.

At the Human Nutrition Research Center on Aging in Boston, we have been studying many of these declining functions in an effort to see what the relationship is between nutritional phenomena and aging. We would like to be able to influence the recommendations about diet. These are perhaps most widely promulgated in what we call the Recommended Dietary Allowances, which are released every 5 years by the National Academy of Sciences' National Research Council. The 1989 version, as do the previous Recommended Dietary Allowances, describes recommendations for adults between the ages of 23 and 50, and 51-plus.

Not only is that age grouping a total underestimation of the heterogeneity among individuals who are more than 51 years of age, but the amounts recommended are not going to be any different in that category from those in the younger category. So at least we can thank the committee for their honesty in not putting in numbers where insufficient data exist.

Before 1994 or 1995 when the next dietary allowances are published, we must obtain data and information from a variety of sources, including human studies that address the issue of the metabolic behavior, the metabolic needs, and requirements of the increasing segment of our population.

On the basis of some of the research conducted at our center, we have found that the decline in muscle mass over decades can be influenced significantly by physical activity. Even in individuals 90 years old, we are able to show that previously sedentary individuals put on an exercise regimen will increase their lean body mass and significantly increase their muscle function. We know also that if those individuals exercise, their protein needs will be altered, because the efficiency with which protein is used is somewhat lower in the elderly. Therefore, we may have to examine protein requirements in the elderly as compared with requirements in younger individuals. Our findings will be tempered by the ultimate observation that the elderly use protein in such a manner that their kidneys are stressed somewhat if there is too much protein in the diet.

We know that the elderly utilize vitamin A differently; vitamin A is disposed of more slowly. A supplement which would be well utilized by younger individuals may cause stressful circumstances in the elderly.

We know that the elderly do not form vitamin D in their skin as efficiently in the presence of sunlight as do younger individuals, and therefore may be at a greater risk of vitamin D deficiency. Thus, they may need more vitamin D in their diets than younger individuals.

We know that elderly individuals make less stomach acid and that the acuity of smell diminishes. All of these factors will influence the kinds of recommendations and eating behavior that we will address as we examine both the nutritional needs and the physical activity of older individuals.

Our research about metabolism and function in the elderly can perhaps influence the next publication of the Recommended Dietary Allowances. If we are to have the flexibility and make the kinds of decisions needed, we must generate a great deal more information, and we must find ways of integrating that information into the food system. The only way to accomplish this task effectively in the next decade or two will be through more effective interaction among industry, academic institutions, and government. What the model should be for that interaction is not clear, but it will ultimately be of enormous help not only to business, but to the health of the population.

BACKGROUND REFERENCES

Bortz WM. Disuse and aging. *JAMA* 1982;248:1203.

Fiatarone JA, Marks EC, Ryan ND, Meredity CN, Lipsitz LA, Evans WJ. High intensity strength training in nonagenarians. *JAMA* 1990;263:3029–34.

McGandy RB, Russell RM, Hartz SC, Jacob RA, Peters H, Sahyoun N. Nutritional status survey of healthy noninstitutionalized elderly: nutrient intakes from three-day diet records and nutrient supplements. *Nutr Res* 1986;6:785–98.

Meredith CN, Zackin MJ, Frontera WR, Evans WJ. Dietary protein requirements and body protein metabolism in endurance-trained men. *J Appl Physiol* 1989;66:2850–6.

Munro FN, McGandy RB, Hartz SC, Russell RM, Jacob RA, Otradovec CL. Protein nutriture of a group of free-living elderly. *Am J Clin Nutr* 1987;6:586–92.

Schock NW. Some physiological aspects of aging in man. *Bull NY Acad Med* 1956;32:268–78.

Webb AR, Kline L, Holick MF. Influence of season and latitude on the cutaneous synthesis of vitamin D_3: exposure to winter sunlight in Boston and Edmonton will not promote vitamin D_3 synthesis in human skin. *J Clin Endocrinol Metab* 1988;61:373–8.

Zheng JJ, Rosenberg IH. What is the nutritional status of the elderly? *Geriatrics* 1989;44:57–64.

CHAPTER 13

Nutrition and the Consumer

Judith S. Stern

Today there is no one consumer. The segmentation of the marketplace is best seen by walking down the cereal aisle in a supermarket. Health-conscious consumers may choose cereal with oat bran, which may reduce blood cholesterol when consumed along with a low-fat diet. Heartwise, a cereal made with the soluble fiber found in the laxative Metamucil®, is targeted at the consumer who is 40 years and older. For these consumers, nutrition, not taste, is often the primary factor. It may not taste terrible, but it probably will not make the top 10 favorite food list.

At the other end of the spectrum is the consumer, who is not really interested in nutrition, but looks to certain foods for their entertainment value. Teenage Mutant Ninja Turtle Cereal is described as having Ninjanets, with marshmallow Mutant Turtles, and you can write away for Nintendo video games. There is even a Barbie Doll Cereal. No value judgments are made about the nutrition pluses and minuses of these cereals; they are described to illustrate the variety of consumers.

The consumer in the middle is the individual who cares somewhat about nutrition and health and if the message is strong enough and unified enough, may modify his or her food intake or food choices. Thirty percent of the 3.6 billion dollar U.S. food advertising budget contains a health message and may convince this consumer to buy a specific product based on health.

Prediction of consumer demands for the year 2001 should take into account the current segmentation of the marketplace and try to predict future segmentation. I believe that the number of health-conscious consumers will increase in the next 10 to 20 years and the increase will help fuel the demand for "medically designed" foods. My contention is based on a number of factors, including increasing consumer interest in nutrition and the emergence of a number of reports in which there is a consensus about the link of diet and health.

An annual telephone survey of over 1000 primary food shoppers done by the Food Marketing Institute asked shoppers how they rate the importance of various factors in food selection (Table 1). Taste was cited as the most important factor. If it tastes terrible, people won't buy it again. Seventy-six percent of these consumers say that nutrition is very important. This is followed by product safety (74%), price (64%). This latter figure might be different if lower socioeconomic consumers were surveyed.

Table 1   Importance of Various Factors in Food Selection

	Percentage Responding "Very Important"
Taste	87
Nutrition	76
Product safety	74
Price	64
Storability	40
Preparation time	37

Source: Food Marketing Institute (FMI), 1989.

According to figures in a previous report, the median age of our population will increase within the next 10 to 20 years. We know that older consumers are more interested in health than younger consumers. Readers of magazines that emphasize nutrition and health tend to be older. Designing foods for these older consumers will mean learning more about their nutrient needs. We have begun to establish RDAs for older people (beyond 51 years of age). Marketing foods to older individuals will necessitate different marketing approaches, including accounting for segmentation within this older consumer group.

For the first time, there is consensus in the government (based on their publications) as to what constitutes a healthful diet. Although we may modify some of these recommendations in the next 10 years, nonetheless a consensus exists. Included in the consensus are the Surgeon General's report in 1988, the diet and health report in 1989, the implementation of the dietary guidelines report that, it is hoped, will be published by the end of 1990, and the report on nutrition labeling, which also will be published by the end of 1990. All of these reports support the recommendations that many Americans should consume a diet lower in fat.

There may have to be additional legislation to help the food industry meet consumer needs for products lower in fat. This may require changes in "standards of identity." For example, if a low-fat cheese is available, it would be helpful in marketing not to have to label it "imitation cheese." There are now changes occurring in the USDA system of grading meat. We now have a grade called "select" that is lower in fat than "choice" or "prime." I hope the USDA will continue to modify the grading system to put a premium on lower fat cuts.

Some of the continuing interest in nutrition is undoubtedly media driven. The media will continue to be active in this area. The October 12, 1989, issue of *The Wall Street Journal* had the following articles on page B1: "A Fig Leaf a Day Keeps the Doctor Away" (an article about using fig leaves for new sources of dietary fiber); "Dairies are Skimming Cholesterol from Milk" (this development in technology reflects consumer pressure for foods lower in cholesterol); and finally, "Biodegradable Traps to Boost Lobster Catch" (this reflects environmental issues).

That same week in *The New York Times* there was an article entitled: "Snacking is Said to Cut Heart Risk." The latter study supports modified dietary patterns to improve health. We are already becoming a "grazing" society in that we are eating more snack meals and fewer large meals. There is opportunity for the food industry to develop new products in this area.

With this interest in nutrition, are consumers making more nutritious choices? When consumers were asked, "What are you eating more or less of to insure that your diet is healthy?" They responded—more fruits and vegetables and less red meat (Table 2). Does this lead to a better diet on the part of consumers? In the case of middle class women, who presumably are literate, but not necessarily scientifically literate, it does not. Based on a recent USDA survey, women who are eating less red meat (presumably to cut back on fat in their diet) are actually consuming more fat in their diet than lower class women. This is because when they decrease the amount of red meat, they eat more salads with high-fat salad dressings, more high-fat cheese, and more high fat frozen desserts. They are also getting less iron; meat is a good source of readily absorbable iron.

Let us now turn to the question of food safety. "How safe is my food?" The consumer perception is that food is not safe, and if you eat it, you will die. When some consumers shop, they are often concerned with avoiding certain things in their food, such as pesticides and food additives. They shop not on the basis of a rational system of choosing a good diet, but instead on the basis of what to avoid.

There is a general fear of chemicals. If we had to put an ingredient label on a "natural" product such as beef, the list of the natural constituents could strike fear into a scientifically illiterate consumer (Table 3).

Part of the problem is that consumers do not understand the concept of "relative" risk. When you ask consumers, "On whom do you rely *most* to be sure that the products you buy are safe?," 41% responded myself (Table 4).

Only 23% responded the government.

This is inappropriate, given the fact that consumers are very far removed from the origin of food. How can the consumer monitor the level of aflatoxin in peanuts for example? You cannot see it, you cannot smell it. In the area of aflatoxin contamination, one must rely on the government to be sure that food is safe. "Relative" risk has been described using the example of Alar in apples. If there is a risk associated with Alar in apples, it is an extremely low risk. Nonetheless, when this made the headlines in April, 1989, a mother, hearing this on the radio, was so concerned that she dashed to her car, exceeded the speed limit, cut off a school bus, and took the apple from her child's lunch. I would contend that speeding and cutting off the school bus posed a greater risk than Alar in apples.

In 1989, consumers avoided Alar in apples (extremely low risk) but "forgot" to wear seat belts (moderate risk), smoked cigarettes (high risk), or went hang gliding in California (very high risk). As stated by Dr. Frank Young of the FDA, the question asked about food should not be, "Is it absolutely safe?," but "Is the margin of safety adequate to protect the consumer?"

In summary, if the consumer were asked today what she or he would like in future foods, the answer might be: cure obesity, prevent cancer and heart disease, be absolutely safe, be in biodegradable packages, have nutrition labels that list everything, and taste terrific. These responses reflect public awareness or public naivete, but they are not consistent with the science base or with what we can do technologically.

Table 2   What Are You Eating More or Less of to Ensure That Your Diet Is Healthy?

	Percentage
More fruits/vegetables	59
Less meat/red meat	33
Less fats/oils	22
Less sugar	20
More fish	18
More chicken	16
Less salt	13

Source: FMI, 1989.

Table 3   Ingredient Labeling of Beef by "Chemicals"

Contents: Water, triglyceride of stearic, palmitic, oleic, and linoleic acids, myosin and actin (containing glutamic acid, lysine, leucine, arginine, aspartic acid, alanine, serine, isole-ucine, phenylalanine, theonine, valine, tyrosine, proline, histidine, methionine, cysteine, and tryptophan in peptide linkages), glycogen, collagen, lecithin, cholesterol, phosphocrea-tine, carnitine, dipotassium phosphate, myoglobin, and urea.

WARNING: May contain steroid hormones of natural origin.

Table 4   On Whom Do You Rely *Most* to Be Sure That the Products You Buy Are Safe?

Yourself	41%
Government	23%
Manufacturers	14%
Retailers	10%
Consumer organizations	8%

Source: FMI, 1989.

BACKGROUND REFERENCES

Food Marketing Institute. *Trends: Consumer Attitudes and the Supermarket.* Washington, DC, Food Marketing Institute, 1989.

Lekoski L. Cereal. *Supermarket Business* 1989;176:168–9.

Ritson C, Gofton L, McKenzie J, eds. *The Food Consumer.* New York: John Wiley and Sons, 1986.

PART IV

FOOD, NUTRITION, AND HEALTH: THE ROLE OF GOVERNMENT

ABSTRACT

Secretary of Agriculture Clayton Yeutter focuses on the politics of food and discusses the public's fear of biotechnology. He blames public advocacy groups for using the media to turn the people against the new technology. In the face of their tactics we all have a responsibility to do a better job of explaining the important role of biotechnology in feeding the world more cheaply and efficiently. As a case study Secretary Yeutter discusses bovine somatotropin, or BST, the hormone that promises to increase dramatically the efficiency of diary production throughout the world. He questions the scientific basis for attacks on BST by advocacy groups and disputes the claim made by dairy producers that BST would put them out of business by generating huge milk surpluses. In Secretary Yeutter's view everyone seems to be overlooking the fact that BST will mean lower food costs for consumers. He warns that all the recent criticism of biotechnology will cause researchers and especially industry executives to throw up their hands and say: I've had enough of this! If that were to occur, the loss to humankind would be tragic.

Consumers are bombarded by so many conflicting health claims on food packages and in the media that they do not know what to believe. In the end, many are driven to reject the basic idea that there is a link between diet and health. Recent surveys indicate that the majority of Americans are not concerned with nutrition. A recent Gallup Poll reported that only 32% of those surveyed knew their blood cholesterol level. Secretary of Health and Human Services Louis Sullivan points to the pressing need to convince consumers of the link between diet and health. In his view, food labeling is of paramount importance in educating consumers. At present, 60% of packaged foods regulated by the FDA have nutritional labeling. Sullivan suggests that there should be a higher percentage so labeled. In any case, we must make certain that labels are accurate, honest and easy to understand.

Nowhere is the interaction of scientists and policy makers more critical than in the field of nutrition. Throughout the 20th century, nutritional research has resulted in important policy decisions like the iodization of salt, vitamin D fortification of milk, fluoridation of water, and food safety regulations.

Frank Young, the Commissioner of the Food and Drug Administration, argues that neither our universities nor the media have succeeded in educating the American public about genetic engineering. Dr. Young warns that if we are not careful we shall be forever fending off crises like the public's misplaced fear of Alar in apples. On the question of BST, he reminds us that the FDA has determined that the milk from BST-treated cows presents no safety problems for human beings.

Young identifies several key questions about the future regulation of biotechnology. Do we need to limit the number of genes that can be introduced into a particular type of plant of animal? How many new genes can you add without compromising safety? A single plant breeder may introduce so many new genotypes per year into a field that regulating plant by plant is out of the question. We need to develop broad categories for scrutiny of new products.

Sanford Miller, former Director of the Food and Drug Administration's Center for Food Safety and Applied Nutrition, offers his perspective on the role of regulators in the age of biotechnology. According to Miller, the biggest mistake that industry can make is to assume that weak regulatory agencies are in its interest. On the contrary, weak regulators will always say "no," because they are afraid to make judgments. Miller also stresses the importance of international cooperation in setting food safety standards.

Jack Moore, former Deputy Administrator of the Environmental Protection Agency, warns that if we do not solve the problem of a scientifically illiterate public in the United States, biotechnology will amount to nothing. The public feels that science has visited awful things upon them, like cancer and birth defects. Scientific arrogance—telling the public that we know we are right so just believe us—is not going to solve anything. Moore urges us to invite the public into the scientific dialogue and let them know what genetic engineering is about and what it has to offer them.

CHAPTER 14

The Politics of Food and Biotechnophobia

Clayton Yeutter

We are in the throes of vast changes in American and global agriculture. We no longer can separate our agricultural plant or marketing system from that of the rest of the world and, therefore, it is most appropriate that people from various countries throughout the world participate in a conference that looks toward the 21st century.

The U.S. Department of Agriculture is intensely involved in biotechnological studies and in agricultural research generally. In biotechnology alone we have approximately 1000 scientists scattered through the system in the United States who are working on biotechnological issues. Another 1000 postgraduate students are included in that effort, and controlled field tests are being conducted on a host of products in a number of places throughout the country. There are equally intense efforts underway in the private sector, here and elsewhere in the world.

My impression from talking with groups since I became the Secretary of Agriculture is that most Americans still do not fully understand the magnitude of what is involved in biotechnological research. The term is recognized but not well understood.

The potential for humankind in biotechnological research provokes awe, both in terms of human medicine and in terms of agricultural production. All of us have a challenge and a responsibility to do a better job of explaining precisely what is at stake in this research effort and what it means to all of mankind. If we consider agriculture as an example, and look at countries around the world, we realize that biotechnology could have an impact on agricultural production in the Soviet Union. The Soviet Union has an unfriendly environment for the production of food, but that is changing for the better. Certainly, biotechnology can provide some answers for improved food production in the Soviet Union and elsewhere.

Consider for a moment Brazil's acidic soils. Biotechnology may provide advances in crop adaptability to the kinds of soils that have affected food production in Brazil. If one looks at the magnitude of potential production in that part of the world, it is easy to perceive the importance of advances in biotechnology.

The same considerations apply to studies in the United States. Even though we are perhaps the most efficient food producer in the world in most products, it is important for us to stay ahead of the game. Biotechnology offers us the potential to maintain our technological edge over competitors throughout the world.

Technology and management separate American agriculture from the agriculture of most other countries throughout the world. Our advantage can be maintained only by expending effort. It is not that we have lower labor costs or, for that matter, lower land costs than most of our competitors. Our advantage can be maintained only if we sustain our edge in technology and management.

We do, however, face some threats in the coming months and years. One, of course, relates to the question of health concerns—food safety. The pendulum of concern seems to have swung too far to an extreme, and it is incumbent upon all of us to enable people to examine these issues with greater objectivity.

Perhaps the best example at the moment is the concern over the potential introduction of bovine somatotropin (BST), the hormonal product that promises significant increases in dairy production throughout the world. Well before the scientific evaluation of the product by the Food and Drug Administration is complete, the product is the focus of criticism, both with respect to its potential effect on the dairy cows injected with the product and with the possible health effects on human beings who consume dairy products produced from animals injected with BST. To my knowledge, however, there are no legitimate studies thus far to indicate concern in either area. Why has BST become such a producer of grave and troublesome headlines?

Without question many of those attacks are at least premature and perhaps irresponsible. A double standard seems to have been applied to the food safety issues by the media. A recent editorial suggested that the standard by which scientific evidence provided by government, that is, USDA, FDA, and EPA, appears to be a totally different standard from that applied to the advocacy groups who are attacking products on the basis of alleged food safety concerns. Certainly, scientific evaluations within the Department of Agriculture or any other governmental agency should be subjected to public scrutiny and the pressures of careful evaluations by those outside of government, but equal standards should be applied to those who levy the attacks. The media should judge our claims and those of our detractors by the same criteria. This does not always appear to be the practice.

Concerns for food safety also have affected those who sell the products, or sell the food products whose production process included some of the new agricultural chemicals such as BST. Alar in apples was the first example (and more recently BST) of an agricultural chemical that prompted advocacy groups to attempt to obtain commitments from food retailers not to sell apples with Alar or milk from animals injected with BST. Surveys completed in the aftermath of the Alar controversy, for example, indicated that the American public had not lost confidence in apples as a result of the use of Alar. People did not stop buying apples because of Alar; they stopped buying apples because the supermarkets were not carrying them.

The retailers took apples out of the marketplace because of the fear of the impact of all of the adverse publicity and the potential economic price incurred by loss of customers. We must work to avoid intimidation by those who have a particular axe to grind in the area of food safety. If apples are unsafe, retailers should not put them on the shelves, and people should not buy them. Scientists, however, should provide those answers. The art of intimidation has no place in a democratic society.

The economic implications of these kinds of issues affect all of us. Concerns about the potential use of a product such as BST have been expressed by the dairy industry, that is, by the milk producers. They fear that a substantial increase in milk production will generate surpluses that will then cause budgetary problems for the industry and for the Department of Agriculture, the agency that must remove those surpluses from the market.

The comment that I have heard from people in the dairy industry is, "Why should we use a product that nobody wants or needs to produce a product that nobody wants to buy?" In other words, why should we generate surpluses? When that kind of argument emerges, all its facets must be examined with sound judgment and science. The examination thus far of the economic implications of the BST controversy has been greatly simplified. The issues are more complex than they appear.

For example, we appear to approach the issue of BST with the economic mentality that has prevailed on agricultural policy issues in the United States for much of the last 50 years. That is, a mentality characterized by an emphasis on the inelasticity of demand for food products, that is, if we produce more, we earn less because of the inelasticity of demand.

Should we assume, however, that demand for American food products is inelastic? If we consider only a domestic marketplace with milk or any other domestic product, the assumption probably will be proved correct. Our demand for most foods in the United States is relatively inelastic.

Today, however, we must consider global demands. If we are truly in an international marketplace, we must evaluate demand on a worldwide basis, and not just on that of the domestic marketplace. The demand internationally for most food products, for most American exporters at least, is elastic, not inelastic. If we were internationally competitive in dairy production, and perhaps we could be through the use of a product such as BST or another product that would confer a temporary technological edge, we would take a much different view of dairy production with BST. We would assuredly sell dairy products at a lower price, and we might sell much more than we do today. Therefore, the incomes of milk producers would rise, not fall, with the application of such a product.

I see no discussion whatsoever of the potential elasticity of demand for American dairy products within the dairy industry. I regard the lack of discussion regrettable indeed. The potential benefits to consumers of lower food costs appear to have been lost in this debate. Where are the consumer advocacy groups? Why is Ralph Nader not focusing on this issue in regard to food costs? Are consumer advocates concerned about low-income consumers? Milk is one of our most complete food products, perhaps the most complete product. Enormous potential benefit would be conferred on low-income families in the United States and throughout the world if a product, such as BST, were to be developed to lower the cost of milk. Those who spoke vigorously on behalf of consumers in the 1970s when food costs were rising are quiet today as we evaluate issues such as this. I have heard nothing from any of the regular spokespersons for consumer interests.

One of the great potential benefits of biotechnological research is that it is one means by which inflation in the United States and in other countries throughout the world might be controlled. Nevertheless, no discussion of that potentially positive economic impact has occurred.

One ultimately comes to the issue of international competitiveness. Where are the arguments about the benefits of technological advantage here in the United States? If we can gain an edge in technology through this research effort, the research should be conducted.

One final comment on this topic. What is the potential impact of the criticism on the researchers themselves? If a scientist is engaged in biotechnological research and produces a product such as BST that is severely criticized, what is the consequence? At some point, the researcher may leave the area of research entirely.

Biotechnological research must be demystified for the public. Advocacy groups exist with the objective to eliminate all agricultural chemicals by 1995. That is a serious concern for all of us. I am confident that effort will not succeed because it should not succeed. But the fact is, if we eliminated all agricultural chemical uses by 1995, we would have another major set of problems in this country and in this world. Therefore, the reasons for engaging in biotechnological research must be articulated and advocated with the same vigor as the concerns expressed by advocacy groups.

Within the administration we have been giving a great deal of attention to the question of food safety and the control of agricultural chemical registrations, re-registrations, and suspension activity. As a result we have developed an Administration position on a number of these issues that will soon be announced.

One of our primary objectives in agricultural negotiations lies in the area of sanitary and phytosanitary regulation and rules. We need much greater harmonization of international rules and standards in the entire area of food. Equitable rules regarding food should be applied universally. Different rules for different countries lead to problems.

A classic example has been the European Community ban on American beef produced using hormones. We believe there is no scientific basis for the ban in Europe, and most people in Europe agree with us. If standards were equal in regard to beef production, we would not have this trade conflict and American beef producers would not have lost a $100-million-a-year market. The United States retaliation against $100 million worth of European Community products coming into the United States would not have occurred.

Trade liberalization is also important. We are interested in feeding the world, in raising living standards throughout the world, in responding positively to the challenge of improved nutrition and diet for hundreds of millions, perhaps billions of people throughout the world. Only through trade liberalization can we make a major impact.

Your interest and the interest of people everywhere should be stimulated by the Uruguay multilateral trade negotiations. This is probably the finest opportunity that the world has had in the area of trade liberalization in all of history. The profile of the exercise is not sufficiently high. Those of us who are interested in agriculture and in trade liberalization should be aware and help increase the awareness of others.

I can assure you that unless those who have the breadth of understanding and interest to be concerned about trade liberalization speak up, we will not accomplish our goals.

We must all work to improve public understanding of biotechnological research and the benefits that accrue. We have a common educational challenge. We must do a better job of allaying fears than those who inspire fears of biotechnology, a science used to better the lives of people around the world.

CHAPTER 15

The Link Between Nutrition and Health

Louis W. Sullivan

Intensive research can yield enormous dividends for the food industry and for mankind. We shall enjoy improvements, innovations, and inventions that will give us new and more healthful food choices. Genetic engineering and biotechnology will create miracles to help us feed a hungry world efficiently and economically.

Thus, genetic engineering and biotechnology could help improve the lives of those who live in developing countries, help them reduce their need for aid, and enable them to contribute to the economic vitality of our world. Forecasting, however, must be tempered with fact. We are a long way from talking substantively about nutrition and health in the next century, because we have yet to resolve the fundamental issues that challenge us now. Two related issues must be addressed: first, there is clear and compelling evidence that diet affects our health, and yet some consumers are confused by conflicting information. As a result, they ignore sound nutritional advice. The second issue that concerns me is that of food product labeling. We want to make it easier for consumers to reduce the fat in their diet, to cut their dietary cholesterol, and to increase their fiber intake. Together, we must make certain that product labels are accurate, honest, and easy to understand.

Many consumers readily accept and are concerned about the link between nutrition and health. Commercially, health sells products. In the first half of 1989, 40% of the food products introduced in the marketplace had health messages. Oat-based cereals were selling so well that half the new cereals were oat-based. Manufacturers are creating crackers and cookies they call "cholesterol free." One securities analyst projected that there is tremendous growth potential for healthful food. Health-based marketing is clearly the key to future sales, but there are still too many people who don't believe in the nutrition–health link, or are confused about what is best.

Consumers have a right to be confused. For the public, the latest statement on diet and health appears to be the best evidence regardless of its accuracy. For example, a well-read newspaper recently printed this headline, "Diet Changes Could Spell Arthritis Relief." The story suggested that the 37 million Americans who suffer from arthritis, "may find relief by changing their diet." The story reported a survey in which one-tenth of those interviewed said, "dietary changes helped their condition, in some cases dramatically." More than half of those surveyed said their doctor gave them no nutritional information, or said there was no relationship between diet and arthritis.

Conflicting advice and information about nutrition will lead some to rebel against the link between diet and health. Too many consumers really do not understand nutrition, and among those who do, we have yet to determine the means to translate their awareness into behavioral change.

How may people know the meaning of words like "protein," or "carbohydrate," or even better, "complex carbohydrate." The potential exists to influence some consumers against good nutrition even more readily than they can be influenced in favor of a more healthful diet.

Consider these headlines: from *The Chicago Tribune*, "Medical Experts Downplay Dangers of Cholesterol," and from the *Baltimore Sun*, "Cholesterol Downgraded as Heart Disease Risk Factor." The accompanying stories reported that there are many factors involved in heart disease including smoking and high blood pressure. But one of the stories cited journalist Thomas Moore's misleading *Atlantic Monthly* article, which attempted to argue that studies have failed to establish that diet has an effect on cholesterol and that lowering cholesterol can save lives.

The other story reported, "after years of warnings to Americans to avoid red meat, milk, and eggs, a panel of medical experts said that cholesterol's role in heart disease has been exaggerated and is worrying the public unnecessarily." Anyone looking for an excuse to abandon a low-fat, low-cholesterol diet in favor of a return to saturated fats has just been handed an invitation to disaster. People believe what they want to believe.

On the basis of presently available data, I am convinced that there is a link between blood cholesterol and heart disease. I am also convinced from the available data that high blood pressure and smoking are among the other key risk factors. But we are not winning enough consumers to this line of thinking. For example, a Minnesota market research firm studied the food buying decisions of people over 50 years of age. The researchers concluded that less than half the people studied were concerned with nutrition. Nearly half of those surveyed followed recommendations on increasing fiber and decreasing salt, fat, and cholesterol, but what of this 51% who ignored this potentially life-threatening/life-lengthening guidance.

A Gallup Poll early in 1989 reported that only 32% of those surveyed knew their blood cholesterol level. Progress is being made, however, because one year earlier only 17% of those surveyed knew their cholesterol level. This increase is encouraging, and it is consistent with survey results from the Department of Health and Human Services. Nevertheless, the evidence shows that only one-third of the population is concerned about cholesterol levels.

Our National Cholesterol Education Program has a monumental task. The food industry has a vested interest in consumer education and must help convince the public that a proper diet will lengthen their lives, reduce health care costs, and increase the quality of their lives.

In time, consumers will be better informed through improved food labeling. For those who are nutritionally aware, and for those who are not, changes must be considered in product labeling. At present, 60% of packaged foods regulated by the FDA have nutritional labeling. Is this sufficient? Should it be extended to all packaged goods? How can consumer confusion be reduced and how can consumers be convinced that a better diet over a lifetime will improve their health?

To answer these questions, the Food and Drug Administration held a series of public hearings to help expand interest in the nutritional value of foods and the effects on diet and health. The Institute of Medicine will provide its best assessment of how to improve food labels. The FDA will attempt to determine what we can do to serve the nutritionally aware consumer and the consumer who is not yet convinced that diet makes a difference.

Consumer responses to the following issues will help guide government agencies concerned with nutrition and health:

What policies should FDA apply to regulate health claims on food labels that say a particular product may treat or prevent disease?

Should nutrition labeling, which is now voluntary unless a nutrition claim is made, be required for most or all packaged food?

How can labels help consumers identify foods low in saturated fat, cholesterol, sodium, and high in fiber?

Should the type of oils and fats used in a product be listed?

Do descriptions such as "low fat," "high fiber," and "light" need definition?

Answers to these questions will enable health and human services to educate those who are nutritionally aware. Confused consumers and skeptics must be influenced to accept the critical link between diet and health. From planting to processing, from packaging to promotion, the food industry is intimately affected by consumer decisions. The more we know about those decisions and how to influence them for better health decisions, the easier it will be to plan for the future. In the final analysis, a healthier future is what nutrition is all about.

CHAPTER 16

Regulatory Challenges and Biotechnological Advances

Frank E. Young

A s a nation, as we deal with foods, and we must make policy decisions in the years ahead, a prerequisite for sound decision- making in regard to technology and science is that the decision- makers and the public be educated. We must be concerned with (1) how nontariff barriers are likely to affect the food industry, (2) the effect of environmental impact analyses, and (3) how we deal with risk assessment in new areas.

I will use as a paradigm some areas in biotechnology that I think will markedly change the face of the agrifood business. Specifically, I will focus on bovine somatotropin (BST), which is a naturally occurring hormone produced and present within the body of all cows. When eaten by man, it is cleaved into inactive fragments in the digestive tract. Somatotropin has always been present in bovine milk.

There are currently four new applications for marketing approval of BST before the Food and Drug Administration (FDA). The agency has already determined that meat and dairy products from cows so treated pose no safety problems for humans, but we must continue to examine the environmental impact, the safety of the cows, and the efficacy of BST usage. At least two dozen herds are being investigated, and their milk and meat have been introduced into the market during the past several years.

How does the issue of BST-treated animals affect nontariff trade barriers? We can expect to have a wide variety of such barriers as we begin to deal with products developed through biotechnology. One usually discusses trade barriers in Europe and in Japan; we rarely look at trade barriers within the United States.

We appear to have run into a nontariff trade barrier here in the United States when milk from BST-treated herds was boycotted by Ben and Jerry's Ice Cream and by five supermarket chains. BST milk was rejected without a substantial analysis of potential risks and benefits, in spite of studies to show that products from BST-treated animals are safe for human consumption.

The European Community (EC) has placed a moratorium on BST until the end of 1990. Moreover, the EC has recently come up with a fascinating innovation, the so-called "fourth criterion," which deals with social and economic implications of a product's approval.

As we consider BST, it not only has some impact on efficiency of production, but for those of us who are fermentation biologists, it also offers the opportunity to use cattle as a mainstream fermenter.

In regard to environmental impact analyses, we have interesting questions. We have to look at the possible destruction of plants and other dangers to the environment from BST-injected cattle. This is an intriguing problem for the FDA, and we are working with colleagues at the Environmental Protection Agency (EPA) and the U.S. Department of Agriculture (USDA) to define the questions we must ask about large-scale release of cows treated with BST hormone into the environment.

Incidentally, when the FDA considers certain food plants, unlike drugs, parts of premarket investigation ("field trials") are approved by either the EPA or the USDA, and the final evaluation of the product is done by the FDA. The task is daunting, at best, for an agency that tries to cover this $570 billion industry, or 25 cents of every dollar that you and I spend, with 7,400 employees and a budget of about $542 million.

In regard to environmental impact, however, it appears that the use of BST can decrease the amount of manure added to the ground. It is a possibility that has not been adequately explored, but one in which there is much interest. If nitrogen metabolism is increased and, therefore, nitrogen utilization is increased, would there be a decrease in manure production and, as a result, a decrease in some of its adverse environmental aspects?

In considerations of policy, how do we determine what needs to be regulated? In this regard, I believe a fundamental difference exists between EPA and FDA. The approach used by the EPA is to explore the entire regulatory universe and then expedite the review of certain products. The FDA has attempted first to determine what can be exempted and then focus on that portion which needs to be regulated. We will explore within the coming year exactly what does and does not need to be regulated, and the selection process must be completed in approximately 2 years.

A large number of genetically engineered plants has already been introduced into the environment with the consent of the USDA. We must ask how many genes, conferring what degree of change, may be introduced into a plant or an animal before it ceases to be a generally recognized as safe (GRAS) plant or animal? Such questions will lead to substantive debates.

A single plant breeder may introduce as many as 20,000 new genotypes per year into the field. If we attempted to evaluate one plant at a time, the delays in the approval process would be enormous. Broad categories will have to be developed to facilitate the process.

One may ask, why does the FDA regulate at all? There are two fundamental categories we monitor. The first is the production of a new food additive, such as a color additive, or the cloning of a gene, which produces a new protein. This poses the problem of determining when a change becomes a food additive. How much of a change is required for the FDA to become involved?

A difficult question most recently was created by the introduction of a noxious compound or adulterant such as a poison. The FDA must act quickly and prudently on premarket approval in cases of adulteration. Seven of the thirteen vitally important drugs that the FDA has approved since 1986 are new biotechnology products. They have made a major impact on one part of the economy, and we're looking with interest at what is happening in the agricultural and environmental sector, where there has been marked apathy in new products investment.

Another problem that we must address is what happens when you introduce noxious or potentially noxious compounds into a food. Suppose a naturally occurring pesticide gene is amplified so that it produces large amounts of pesticide in the fruit. What have you created? Is it simply a variant of natural food, is it a food additive, or is it a naturally occurring toxin that produces an adulteration in the food? We have less than 2 years to clarify these policy questions. How are we going to do it?

We will discuss with the FDA advisory committees, the National Academy of Sciences, in Congress, in administrative groups of the EPA, FDA, and USDA, the concept of when and what must be done, if anything, about premarket approval of new biotechnological products for human consumption. We must have input from our consumer, academic, and industrial groups, or we will fail.

One of the beneficial aspects of the discussion about recombinant DNA was that they did precipitate the 1976 guidelines, and we had a vigorous debate that was both scholarly and forceful, weighing all sides of the question. We have not had that same debate in the agricultural and environmental realms, nor were the experiments designed to answer some of the risk problems that we saw, such as, the polyoma virus cloned into E. coli vectors and vehicles, whether or not tumors were produced in animals, and how much growth hormone could survive in the intestinal tract.

The media and our universities have been unsuccessful in their attempts to introduce the benefits of genetic engineering into the minds of the American public.

There are three distinct modes of genetic engineering that will be employed in the food industry: one is classical modification of the whole organism. It has been the standard approach for many years. Second, cellular modifications, such as cell fusion after which cellular products are propagated. And the third, the most recent method, molecular modification, for example, in which recombinant DNA technology is used.

At the last Organization for Economic Cooperation and Development (OECD) meeting, there was substantial argument as to what is a genetically modified organism. The regulations in Europe are built around terms that scientists use rather casually and that can have different meanings. Precise use of scientific terms is essential.

What exactly is a genetically modified organism? Is it an organism that has been changed by conjugation or transformation or cell fusion or is it any willy-nilly mutation that occurs from the ultraviolet rays that we absorb? Such definitions will have a huge impact in both Europe and the United States.

Those of us who communicate with educators, government officials, industry representatives, and individuals from various consumer groups, have an enormous responsibility to be precise and careful in what we say. We have recently experienced the Alar in apples and the cyanide in Chilean grapes episodes. The rhetoric ultimately caused great problems for the apple industry. How can such incidents be avoided?

One answer is a well-designed, nationwide education program that would include not only high schools and colleges, but also the electronic and print media. The campaign would require the development of a new media forum that does not exist today. The 30-second to 3-minute spots on television are an inappropriate, unsuccessful means by which to educate the public. I lament the one-liners that I have to deliver from time to time.

Our major challenges are establishing policy decisions and disseminating scientific information to our nation and around the world.

We should debate matters such as the need for an international science court. Should there be an international academic society? How do we deal with the development of new knowledge and translate that knowledge into policy? The primary issue we must address, however, is how to guide and inform the public, legitimately and honestly, in regard to risks and benefits brought by modern science.

Of relevance to the issue of an educated, informed public is an interesting dichotomy. Those who believe that the treatment of diseases such as cancer and AIDS must be expedited propose that time pressures outweigh risks in using new biotechnologically produced drugs. At the same time, ironically, the attitude toward the application of food biotechnology is that nothing that is genetically manipulated should be tolerated. To see these two attitudes played out side by side has been a remarkable experience. Thus, an informed choice can be made only on the basis of equally good understanding of the risks and benefits involved. Such understanding, coupled with declared technocratic policies, will be the stuff of the great debate of the 1990s.

CHAPTER 17

Food Regulation: Future Issues

Sanford Miller

Although the future of regulatory activities cannot be predicted with certainty, there are factors that can be identified as having an impact on the nature of regulation in the next century. Competent scientists are essential to the process. A major mistake made by industry is to assume that it is in their best interest to have weak regulatory agencies. Weak regulators are always afraid to make judgments, but strong, competent scientists in regulatory positions are always willing to make judgments based on their scientific expertise and to defend those judgments. A competent decision is best made in response to a clear mandate from Congress.

The conflict between various interests in the Congress and between various economic and social constituencies affect regulators in their decision-making processes and create difficulties. Future regulatory activities can be divided into four broad areas: first, the changes in technology and potential new hazards that result from the technologies. Equally important are the old hazards that do not disappear. The second category is an understanding of health and safety. The third category encompasses socioeconomic changes. Finally there are international concerns to which we have paid increasing attention over the past decade, but have yet to find a process to ensure international regulatory harmony.

Among the old and potential new hazards are chemicals in our food supply, pesticides, food additives, and packaging materials. The public, rather than the regulators, believe that these substances are major contributors to the hazards in food. Continued public concern and the requirement for premarket approval and monitoring of these substances will continue to make them issues of significance.

Microbiological hazards will continue to grow. This is a continual problem that appears to cycle. The oldest known hazards in our food supply were microbiological, and they continue to be the most important contributors to food-borne disease. As our knowledge of microbiology expands, new organisms emerge as food-borne pathogens. Thus far, the best or most efficient techniques for controlling these organisms have not been determined.

We once believed that refrigeration was our last line of defense against microbial growth. We now know there are psychrophilic organisms, organisms that grow in the cold, such as *Listeria* and *Yersinia*, which appear to be more common in foods today than in the past. More importantly, outbreaks of disease associated with these organisms are being described. We need to know much more about the pathogenic processes of these organisms, the processes that cause their transformation from benign to toxigenic, and the factors that regulate their growth. Only then can we develop methods to control them.

There are several factors that have contributed to this problem: increasing automation in industry, changing eating habits, and increasing mobility of our society resulting in increased consumption of food away from the home. Our methods to regulate these areas of food use are not very good, and are certainly not always effective.

Our understanding of health and safety is always changing. Most of our knowledge of health has been descriptive, and yet to use biology as a basis for regulation means that operation proceeds on a case-by-case basis. The broadening dimensions of our food supply, however, have made this mode impossible, and now food safety and health concerns such as nutrition must be viewed differently, that is, in a more comprehensive theoretical framework.

What are some of these broadened areas of concern? We know how to control those substances in food that kill people. Acute toxicity is not a major concern in food safety. We are gaining a better understanding of the relationship between foods and cancer or cardiovascular disease. We are learning to control carcinogens in the food supply or learning whether such control is necessary.

The greatest concerns are those of which we know the least, the nonquantal factors, the graded responses involving the impact of the food supply on immune function, on behavior and on physical performance. One can ask philosophical questions: "Should we be concerned about these issues at all? How much deeper shall we or need we go in our quest to determine safety?" We are beginning to recognize that the impact of diet on health is an aggregate phenomena. If we are to resolve the conflicts in this field, we must think about diet and health as an integrated, multifactorial problem.

The impact of diet must be considered at even the molecular level, but a still finer level of concern is that of the submolecular level of metabolic regulation.

Do inorganic elements act at their primary level by regulating the "phonic" properties of atoms? Do they regulate the frequency of the vibratory period or the transformation of atoms? And, indeed, do these activities have anything to do with regulation of metabolic phenomena.

Much of this thinking arises from the great controversy concerning the impact of nonionizing, low-frequency radiation on human health. Is there an effect or is there not? Such radiation has been reported to change the phonic properties of atoms. If radiation has such an effect, then do dietary components, regulating the environment in which these atoms function, also modify these activities as well. It is a new area in which investigation has not yet occurred.

Toxicology is now functioning as a mechanistic not a phenomenologic science. It is no longer purely descriptive, but must be understood mechanistically. Otherwise, the process will continue by which we evaluate substances and events on a case-by-case basis and generic criteria will not be developed by which to expedite and improve regulation.

The only rational approach we can take in evaluating foods for safety is to identify relative risk. We cannot guarantee absolute food safety, even though complete safety was implied during the 1930s and 1940s and 1950s by regulators. In reality, they regulated with judgment and estimates of relative risk.

There are social and economic factors that have an impact on regulatory decisions, in particular, the media and politics. Media pressure has a tremendous impact on regulatory agencies. Regulators do not like to find themselves in a contretemps in which they are automatically presumed guilty.

A major factor influencing regulatory decisions and those who make the decisions is that the regulators can never say, "I changed my mind." Consider the cyclamate situation. The agency concluded more than 20 years ago that cyclamates were unsafe, based on the data available at the time. In the last several years, however, reevaluation of the data based on contemporary biology has shown that cyclamates are safe. Nevertheless, the FDA knew that if the necessary action were taken to repermit the use of cyclamates in food, the FDA would be attacked from all sides for "caving in to the pressure from the industry" and, as a result, be accused of malfeasance or other high crimes.

This situation is very difficult. Nevertheless, the FDA began the process to repermit the use of cyclamates, because it was scientifically correct to do so. The process still continues, because it was necessary to confirm findings through counsel with the National Academy of Science. Not surprisingly, they recommended more research.

The issue of international regulation focuses on our global food supply. The United States, with all its capacity for growing and producing food, imports more food than it exports, because it is economically important to do so.

Therefore, we must begin to search for processes that will produce harmony within the regulatory activities in international trade. We need a forum in which trade disputes among nations can be resolved, particularly issues of the science of food safety and nutrition.

The most important result of this effort may be the reduction in the use of safety and health considerations as nontariff trade barriers. In the food area, it is the ploy used most commonly to regulate the import of foods that compete with domestic products. The success achieved by some nations will suggest to others that nontariff trade barriers are a useful trade strategy. The United States must insist that such matters be submitted to an international forum for resolution if we are ever to ensure the free flow of food for all the world.

In regard to national health and safety regulation in the 21st century, consider the quotation of Francis Bacon, "Those who will not accept new solutions must expect new evils." Tried and true regulatory strategies and science, over time, will only result in disaster. Remember, regulators are gatekeepers for the introduction of new technologies. If they do not open the gate, new technology cannot and will not be used.

Regulatory agencies around the world are basically the same, that is, very conservative. They follow the same philosophy: when in doubt, don't. And, yet, to allow new technologies, doubt must be resolved. Doubt can only be resolved when we understand better the basic sciences of regulation including toxicology, nutrition, and microbiology.

We must invest more money in research. This message has been heard many times, but it remains absolutely true. The two fundamental areas, nutrition and food safety, that have enormous impact on our personal health, our national health, and our international economies, receive the least support for research from the federal government, private foundations, and industry. This inequity must be resolved. The responsibility belongs not only to the federal government, but also to the food industry as well. It is unfortunate that the cultural dynamic of the food industry has not lent itself to supporting long-term research. The food industry must reexamine itself to determine what is best for its own future. I submit, support for research in nutrition and food safety is not philanthropy, it is survival.

CHAPTER 18

The Need for a Scientifically Literate Public

John Moore

The application of scientific knowledge in the future will be subject to intensive public scrutiny. Changes will be made in the manner in which we apply pesticides. Some of the changes will include a more sophisticated evaluation of the problem. For example, greater than 90% of all pesticides used today do not get to the target site. As a result 90% presents a burden to the environment for which there is essentially no value added. Obviously, engineering the pesticide into a plant would solve some of the problem. Advances in chemical pesticide applications will radically increase the amount of the chemical that gets to the target site.

The nature of pesticides will change. We now have pesticides on the market that exploit a biochemical susceptibility unique to a given plant, as opposed to the old pesticides that were long-lasting, broad-spectrum compounds. These last two characteristics would probably be the kiss of death if someone were now submitting an application to market a pesticide.

Recycling is going to be a major challenge, particularly in this country and in others with burgeoning populations. We are going to have to do something better with water and with waste. Recycling problems apply to the agricultural sector as well as to the industrial sector.

If we are to be successful in dealing with these issues, the solutions must be understood and accepted by the people who are affected. Some have opined that the general population of the United States has become scientifically and technically illiterate. This is onerous; a process is emerging that demands greater knowledge of technical developments from a society ill equipped to assimilate the information.

If we do not educate the public, the facilitation of some of these changes, whether they be biotechnological or the recycling of water, is going to be a grievous task. The public overreaction to Alar did not happen by chance; Alar was predictable, given its misunderstanding of technical issues. I do not know what the next crisis will be, but there will be one. A major component of the problem is a public that believes that these adverse consequences of "new" technologies are being imposed upon them. My suggestion is that whether it be for pesticide use or to implement biotechnology's entry into the economy of this country, we must educate the people of the United States in a manner that recruits their interest and support.

In the past I have used the term scientific arrogance. As a scientist or ex-scientist I believe that we never have the privilege of disregarding the perceptions of the public, as ill-informed or well-informed as they may be.

We must begin a dialogue into which the public is invited. Offer the opportunity to the public to participate in the process. I do not suggest that we make all citizens of America scientists, let alone toxicologists, but we must enable them to build self-confidence in understanding scientific and technical issues. That means that we share a common, practical base of knowledge to which each may add new information. An informed public will, in my opinion, accept sound technical innovation. Let's give them a chance; the current system of hyperbole and rhetoric is not working.

PART V

THE FORCES AFFECTING FOOD: IMPLICATIONS FOR INDUSTRY

ABSTRACT

The economic unification of Europe has stirred American industry's fears of a "Fortress Europe," impregnable to U.S. exports. The U.S. food industry was recently subjected to a European ban on imports of beef from cattle treated with steroid hormones. Sir Roy Denman, the former head of the EC delegation to the United States, insists that trade barriers are not in the European Community's interest.

Denman denies that the European ban on imports of hormone-treated beef was a protectionist ploy. He says that Europe was a victim of the same forces at play in the United States: advocacy groups who convinced consumers and politicians that food additives are death. The European countries did not ban only American beef with steroids, explains Denman, but *any* beef with steroids—including European beef. But, he points out, any country has the right to say what it will eat or drink, provided it imposes the same restrictions on domestic production.

Denman warns that as food technology develops, such problems will happen more and more, unless we find a way of preventing them. He suggests forming international scientific panels to examine the evidence and to determine what should be done under the international trading rules.

Amid public fears of biotechnology and scientific controversy over nutrition, it is the business community that plays a major role in deciding what new products are worth investing in. Michael Miles, CEO of Kraft General Foods, contends that it is up to regulators, manufacturers of technology and primary users to convince consumers of the benefits and safety of biotechnology. In his view processors cannot afford to take the lead in selling new technology to consumers and do not see it as their primary responsibility. In the end the best way to sell the public on the benefits of the new technology will be to produce new products that are safe and beneficial. Consumers do not want to eat technology. They want to eat food that tastes good, is conveniently packaged, and is nutritious.

Sainsbury's R. T. Vyner laments the public's misunderstanding of modern technology. His company is dealing with these concerns by offering a whole range of "green" products, rather than getting rid of existing products at the first sign of consumer fear.

Investment banker Ted Berghorst predicts the ready availability of capital for companies with the right people, products, and proprietary position. The immediate challenge is applying one of the new strategies for private sector funding, as opposed to just government funding, for research and development of the new science and technologies.

Mexican agribusiness leader Eneko de Belausteguigoitia recounts some of the difficulties in the privatization of business in Mexico and calls for the regionalization of the Mexican, Canadian, and American economies.

Urging the international harmonization of food regulation and standards, Francis Gautier, Vice-Chairman of BSN Groupe in France, is optimistic that the EC will be able to accomplish this by 1992. He is less sanguine about the public's response to advanced food technologies and the misuse by governments of health and environmental issues as trade barriers.

CHAPTER 19

European Community Trade Issues Affecting the Food Sector in the 1990s

Sir Roy Denman

Plans for EC harmonization 1992 began in May of 1950, when the French Foreign Minister, Robert Schuman, in words drafted by Jean Monnet, proposed a European Coal and Steel Community. This was the first political attempt to banish forever the vision of another European civil war. We have had two in this century. The last war alone cost 50 million lives. Whatever follies and mistakes we commit in the European Community, we have avoided another civil war.

The Treaty of Rome in 1957 had an economic aim, that is, to take advantage of the prosperity seen in the United States through its huge internal market. The treaty would remove the barriers between member states. The process started with tariffs, and it was finished ahead of schedule in the middle of 1968. But there are many other barriers. In Europe we have frontier formalities, which means that there are lines of trucks at the border and the driver has to wait two hours in the rain to get 50 documents stamped. Imagine how well off you would be if, between every state in the union, you had such a requirement.

What about different technical standards and specifications? Consider tractors as an example. Between the EC member states there are no tariff obstacles. Is there one market? There is not. A tractor can only be permitted in a certain member state if it has a special kind of laminated glass in the cab window. Another tractor has to have a special kind of lubrication for the steering mechanism. Another member state will require a specific distance between the headlights. So there are 12 separate markets for tractors. The same applies to the whole range of technology. The Phillips Company in the Netherlands has 70 engineers who spend every day adjusting the standards for their television sets and electrical appliances for 12 separate member states. Again, imagine the cost of that kind of arrangement in the United States.

We have to have checks at frontiers because of the differences in what we call value-added tax or sales tax in the United States. We will never have a uniform sales tax. But, when there is a difference between 1% and 38%, half the population of Denmark will do its shopping on Friday afternoon in Germany. This is what the economists call a "distortion of trade."

We abolished tariffs in 1968. Why did we wait until recently to tackle these other barriers? The accession of Britain and Ireland and Denmark was looming at the end of 1968, and in our system of government only one big problem can be accommodated at a time. So the absorption of these other member states took time.

The oil price shocks of the 1970s aroused attitudes that were not conducive to the removal of barriers. We knew other barriers existed, but we were trying to tackle them one by one. Thus, every country would say, "You want my barrier abolished? What about yours?"

In 1985, the situation changed when two remarkable people arrived in the Commission. The Commission is our executive branch, and has 17 members, in effect, our Cabinet Secretaries. Jacques Delors, the former Finance Minister of France, is the President. Lord Arthur Cockfield, lawyer and former CEO of the pharmaceutical company, Boots, wrote the white paper on 1992. This remarkable man understood the bottom line in business. He wrote a white paper that, in effect, said if you want a genuinely single European market, here is how to get it. Here are 279 proposals with a name tag and a timetable on each one. Complete this by 1992, and we shall have a market of 320 million people with freedom for businessmen to move, to invest and to deal as easily as between the states of the American union.

Cockfield, an elderly conservative politician who had not dealt much with continental Europe, was sent to put some salt on the tails of the Euro-fanatics. His first question when he arrived in Brussels was "Can one drink the water here?" This did not get an instant hello from those in Brussels. But the history of religion teaches you there is nothing like a convert. He became a convert. He put forward this program and fought it through. His work has changed a great deal in Europe and the world. Is it going to work? You may expect from me some cries of Europhoria, saying it must work because I believe it will. I remember General DeGaulle, not a great believer in the community, on a television program in the 1960s, momentarily endowed with surprising theatrical talent when he said, "You cannot make Europe by leaping on a chair like a goat and crying l'Europe, l'Europe, l'Europe." I would agree with that.

I shall try to establish some solid grounds for thinking the program will work. First of all there are declarations from the European Council, that is our prime ministers or heads of governments, who say the 1992 process is now irreversible. You can say, well, prime ministers will say that kind of thing. They are politicians.

There is an unmistakable wind of change blowing through Europe, and it is the most remarkable change in my generation. All the consumer surveys produced by our firm show that the young French, Germans, and Italians do not think of themselves primarily as nationals but as Europeans. The action is now emerging in Eastern Europe.

Half of the 279 proposals have been passed, including controversial items such as liberalization of capital movements, deregulation of trucking, very difficult for our German friends, and recognition of professional qualifications which even in the United States is not entirely without difficulty. Tremendous arguments took place in regard to money, but an agreement was reached in early 1988 that settled the question of how much we would spend on the Common Agricultural Policy (CAP) for the next 5 years. We were thus freed from spending every one of our top levels meetings arguing about money.

We also have what we call in our jargon the single European act. This in effect is an amendment to our constitution, allowing majority voting directives for the 1992 program and more powers for the European Parliament. That was a difficult exercise. It was a fight between the supporters of Jefferson and Hamilton, which Americans can well understand. Do you want a federation or do you want a collection of sovereign states cooperating together? Well, neither won big, but the result was in the direction of Alexander Hamilton.

What are the results going to be for Europe and for the rest of the world? We commissioned a fairly exhaustive study to about 25 research institutes, which involved correspondence with about 11,000 firms throughout the Community. Although such studies are always subject to error, the results were that if we abolish these barriers in the Community, our GNP should rise by between 5 and 6%, up to $250 billion a year. Put it the other way around, that is the tax we are levying on ourselves because of all these restrictions, imposed on ourselves, by ourselves.

Unemployment should fall by at least 2 million, perhaps 5 million if member states adopt the proper economic policies. The price level should fall on average by about 6% with the economies of long runs. It will not be a painless process. I used to say to American friends in Washington that this was for us as big an adventure as the opening up of the American West, and you know that was not a painless or easy process. But in the end everyone was better off. That, ultimately, is our hope and our faith.

What is going to happen to the rest of the world? There is a lot of talk about Fortress Europe. I think the phrase, Fortress Europe, is silly. You might as well talk, if you read the fine print of the Trade Act of 1988, about Fortress America.

Our Japanese friends will now be amazed to hear talk about Fortress Japan. There will not be a Fortress Europe for the simple reason that it is not in our interest, and self-interest is always the best guide to international behavior. We have 20% of world trade, excluding trade among our member states. To get ourselves into the position where our trading partners throughout the world could retaliate against us would be self defeating.

There is no law that gives us 20 % of world trade. If we turn protectionist, we will become less efficient. We need to compete successfully in world trade, particularly in the field of high technology. To support the R&D required, Europe desperately needs a single market of 320 million people, not small fragmented markets. Otherwise, we sentence ourselves to becoming a collection of developing countries and that is not our ambition.

Let me tackle one more question in this field and then continue to the broader trade implications. Friends in Washington ask me whether or not we are going to consult the Americans on all these directives. I think truth is usually the best answer in discussion among friends, and we say no, we are not. So Americans say, terrible. We say, look, when the Congress passed the Trade Act of 1988, did they consult the Europeans? Of course they did not. Why should they? No country can permit another country to have oversight over its legislation. Congress cannot. We cannot. But that is an argument about form and not substance.

An American firm that comes to Brussels and has a question or suggestion can go to the people in the Berlayment, which is the seat of our bureaucracy, and be heard. A receptive response is in the best interest of any bureaucrat drafting a regulation about the market in which he has not worked. He knows that he has to have some understanding of commercial reality. If he drafts an unsuccessful regulation that is ridiculed in the market, his or her boss will transfer the consequences directly to him or her, so his or her interests are best served by listening carefully to what the practitioners in the market have to say.

Friction will arise in regard to trade issues. The ban imposed by the Europeans on imports of beef, not just from America but from everywhere, certainly generated friction. Some in the United States said that this was a protectionist ploy by the Europeans. That was not the case. Anyone engaged in trade policy thought it was a disaster to have a conflict with our biggest trading partner on something like this. The problem was caused because consumer groups and national parliaments and the European parliament said hormones were bad.

I plead ignorance to the risks or benefits that may arise from the use of hormones, but the voice of the people said we have to take this measure. Now in terms of our international trading obligations I think we acted correctly. Any country has the right to say what it will eat or drink provided it imposes the same restrictions on domestic production as on imports.

Think of the terrible and infamous days of the 18th Amendment, when there was no GATT. America indicated it did not want to import gin, whiskey and wine, and it stopped its own producers from producing it. That was what we did with hormones.

America, for example, has restrictions on the import of certain fatty cheeses from Europe. They can be imported, but they are subject to a kind of refrigeration process for two months, which ruins the flavor.

Certainly the Americans are within their rights. They impose the same restrictions on their producers. That is your sovereign choice. But I cannot help thinking that if this type of restriction is imposed on imports by a number of countries, that ultimately we will need some form of standing international panel that can examine the evidence and recommend international trading rules.

The European Community and the United States differ on CAP. The CAP was one of the corner stones of the European Community. When the French opened their industrial market to the Germans in 1957, they, of course, wanted something in exchange. But we still remain the biggest customer of the American farmer; we purchase 25% of American farm exports each year. We spend less per farmer on support than the federal government spends on supports. In 1986 we spent $23 billion on supporting 11 million farmers, the US government provided $26 billion of support payments for 3 million farmers.

I do not suggest that we are right and the federal government is wrong. Farmers are a very important pressure group. But the European Community has realized it has to reduce support to farmers, not for reasons of idealism; because it costs too much. There is widespread agreement that the problem in world trade and agriculture relates to too much spending on farmers who produce too much in a saturated world market.

Perhaps progress can be made in the Uruguay round next year. Last year the Americans suggested complete abolition of agricultural subsidies. That suggestion, however, would be supremely unpopular in both the OECD and the United States.

We should realize the ultimate in the short term is unlikely, but if we can settle for a substantial and balanced reduction of agricultural subsidies, then the Uruguay round can succeed. Success is important for two reasons: First, the trading world depends on a successful series of negotiations to maintain the fabric of the GATT, which has made possible the biggest expansion of trade in recorded history in the west. Second, the Congress of the United States needs to be shown that the multilateral route can work.

Toward the end of the century Europe probably will have a single currency. If a single trading area is established by the end of 1992, differences in currency and exchange rates become meaningless. If you arrive in one member state today with the equivalent of $1,000, and change your money consecutively into every currency of the other members of the community, you finally have $400. Ultimately, a single currency will be the only practical resolution of the problem.

If by the end of the century one single trading area, one currency, and a largely common economic policy have been established, who then is going to control it? The Commission, which I had the honor to work in and represent for many years, the executive branch of the European Community, could not and should not, because no one elected it. Power has to reside with the elected representatives of the people. We will have to have political control. In other words, the beginning of a European Federation.

This is the Europe that will have to be reckoned with over the next 20 years. We are organizing a great enterprise. We are trying to build a European union.

The historic moment, May 9, 1950, began the process of the unification of Europe. We hope it can be completed in friendship and partnership with a country that Winston Churchill called "The Great Republic," which has for over 40 years maintained the leadership of the West.

CHAPTER 20

The Food Industry and the U.S. Consumer

Michael Miles

Rewarding shareholders is the principal mission of a corporation. In the United States, that means making money, if not this quarter, then at least this year. In a free-market economy making money depends on understanding what consumers want and then supplying them with products that satisfy better than competitors' products, at prices that consumers find acceptable and that earn the supplier a profit and return on its investment.

Although various research studies show that the factors affecting American consumers' choices of food are many and complex, those choices in food can be described fairly simply. First and foremost consumers want food that tastes good. Simply said, people eat what they like and, except in very rare circumstances, they do not eat things they do not like.

Despite research reports, the second most important thing American consumers want in food appears to be convenience. They want quick, easy-to-prepare food that tastes good. This desire for convenience has grown over the years as more and more American women have gone to work outside the home, and it has become so strong that in the last few years this desire has overcome early consumer fears about "radar ranging" themselves and their food. Today over 60% of American homes use microwave ovens and microwaveable products are among the fastest growing in the grocery store.

Also high on consumers' lists of wants in food are health, or nutrition, or food they believe is good for them. Consumers have difficulty defining these wishes, and in many cases consumers are not sure themselves what they mean by "healthy" or "nutritious" food. The kinds of descriptions they use are "wholesome," "natural," "food without artificial ingredients," and certainly "food without dangerous ingredients or residues." We commonly hear, "food like mother used to make."

Until recently American consumers were able to take the wholesomeness, naturalness and safety of American food products for granted. Today, however, because of a complex and increasingly anticipatory interplay of regulatory, political, so-called public interest and media influences, Americans are becoming increasingly concerned and even frightened about the safety of the food supply.

That concern or fear, where fear actually exists, is translated into consumer rejection of foods that involve technology that has been presented to the consumer in mysterious or frightening ways. Thus, consumers reject the concept of irradiated foods. Consumers reject the notion of dairy products from cows treated with BST. Consumers reject products with even traces of Alar, and so on.

In this situation food processors have a very simple choice. They can try to sell products that consumers reject, such as dairy products from BST cows, or they can tell their suppliers, "No BST milk, please. In my products I will use only milk from cows not treated with BST, and I will in turn offer those products to my consumers because they are what consumers want—wholesome, natural, no-additive products."

Having tried on a couple of occasions to sell people products they did not want, let alone products that they rejected outright, I can tell you that it does not work. And since my success as a businessman is to make an attractive profit and return, I am simply not going to spend a lot of the company's money on commercial propositions that I know will not work, certainly not when "work" is defined as providing acceptable profit and return to the shareholders.

In summary, as a food processor, it appears that the adoption of much of the new food technology will depend on the regulators of our industry, the suppliers of the technology, and the primary users of the technology, in many cases the farmers. Farmers supply food manufacturers with many of our raw materials. If new food technology is to succeed, those three forces, regulators, manufacturers of the technology and primary users of the technology, must take a far more active and positive role than they have up to now in convincing consumers of the safety and benefits of that technology.

Most major food processors today are enthusiastic about the beneficial technology that is emerging. We are prepared to invest as partners with the manufacturers and primary users to develop the technology for widespread commercial use. Processors alone, however, simply cannot afford to take the lead in selling this technology to consumers. They do not see this as their primary responsibility.

CHAPTER 21

The Food Industry in Britain

R. T. Vyner

Food safety is a very serious concern in Britain at the moment. When particular scientific issues are misinterpreted, our people are confused. The issue of *Listeria* in England has caused problems with the introduction of ready meals into the National Health Service. England also has the issue of BST. Our government is managing BST in a slightly different manner from that in the United States. They are introducing milk produced from BST- treated cows into the government-controlled milk system, which is causing an outcry, not so much about BST, but rather about the fact that the government is introducing the product into the milk supply system surreptitiously.

Product irradiation has not occurred in Britain. The process has been approved, and it is fascinating to see how quickly fixed positions have been assumed. Some retailers have stated that under no circumstances whatsoever would they handle irradiated foods.

Sainsburys' position is slightly different. We believe the public has a right to a choice, provided we are satisfied as an organization, that is, that our food science and technology and quality control sections are satisfied that irradiation is safe and it is approved by the government authorities. Provided that irradiation controls are in place, we then are committed to give the public a choice.

In England we have the Green movement and food manufacturers must deal with environmental issues. We have our own brand, the Sainsbury's brand, which is the largest single brand in the U.K. We believe that all our brand products must recognize the Green issues, such as the disposal of food packaging and the biodegradability of shopping bags.

Our enormous consumption of plastic shopping bags poses a problem that
we have yet to solve. We need help from science. The current carrier bags that
we use are not biodegradable and are not photodegradable. They are very
strong and all the alternatives we have considered are either far too expensive
or too weak for the job. The biodegradable bag is weaker and costs approximately three times that of our existing bag. We have a commercial problem in
which we have to balance commerce with the desire to recognize the Green
issues.

Global standards and demographic changes affect our business directly and
indirectly. We have a responsibility to provide a respectable profit for our
shareholders and products that consumers are willing to buy. We also must be
cognizant of scientific progress and respond in an objective rather than
subjective manner. Science and business must cooperate for the ultimate
benefit of all involved.

CHAPTER 22

Investment and the New Technologies

D. Theodore Berghorst

I shall provide a brief view of Wall Street's perspective on biotechnology development in the area of food and nutrition and talk about how we finance the development of technology in the food and nutrition areas.

With regard to the Wall Street perspective on biotechnological development in food and nutrition, I believe the applications and future impact of biotechnology on the food and agricultural industries will be substantial. However, Wall Street's perspective today is that such influences are less likely to be revolutionary than incremental. To date, the most visible current biotechnological efforts in the food industry have focused on reducing cost and increasing production capacity. After this initial focus Wall Street will look to the food and nutrition industry to establish additional initiatives to focus on qualitative changes and improvements.

Specific examples are (1) better tasting and better smelling food products; (2) safer foods, with fewer chemical residues from pesticides and other agrichemicals; (3) improved biodegradability and environmental compatibility; (4) new products from grain crops, such as high-fructose corn syrup; (5) the development of more nutritive foods that have a lower content of fat and cholesterol; and 6) additional emphasis on low-calorie foods. In this last category one of the big winners from Wall Street's perspective has been low-calorie beer. Consumer demand for low-calorie beer has driven the use of fermentation in the brewing industry.

The good news from Wall Street is that there are sufficient funds for the development of biotechnology in the food and nutrition areas. Several capital sources are available to support such technological development. Many of the companies in developmental stages are highly financeable if they possess the three "P's," that is, people—good management; product—new products that will be perceived by the consumer as having great benefit; and position—a strong proprietary position. Such investment opportunities, of course, have to be evaluated in the context of strong commercial market potential and have to be "valued" appropriately to interest investors.

The venture capital world today is very strong and very broad with a great amount of capital looking for high-quality investment opportunities. While there indeed has been a strong emphasis in recent years to seek leveraged buyout transactions, nonetheless, there is substantial venture capital available to develop interesting and new technology with high potential.

As the scientific emphasis within such disciplines as microbiology, pharmacology, and molecular biology move out of the world of therapeutics and diagnostics and more directly into the areas of food and nutrition, the venture capital community will be very much interested in investing. The private equity market, which is comprised of major institutional investors and certain types of corporate investors, remains very strong. We are involved in a number of transactions in very early stage companies where resources are available on a worldwide basis.

The public equity market has come back. After the crash of October 1987 through the remainder of 1987 and essentially all of 1988, the public markets were not available to fund emerging new technology companies. The public market for technology situations changed in early 1989. Several companies have successfully completed public offerings this year and there are a number under registration at this time.

Research and development limited partnerships, yet another form of financing, came under some question when the tax law changed in 1986. That has proved not to be an insurmountable problem, and "select" companies today are dropping a "basket" of technology into a legal entity and surrounding it with an R and D limited partnership as a vehicle to fund the technology.

Finally, an area that has grown in prominence, strategic alliance/corporate partnering, is one that I believe makes a lot of sense. In this area emerging technology companies are put together with very large, sophisticated, well-capitalized, major corporations under some type of a technology-transfer arrangement. Those arrangements typically include an up-front technology access fee, an equity infusion in trade for some rights to the technology down the road. Through an ongoing development program milestones are established and the major corporation largely funds the ongoing development of a given technology over a period of 3–5 years. Upon successful conclusion of the technology development a distribution and manufacturing arrangement is put into place. There have been several very successful transactions of this nature. Significant value can be provided by the major company in terms of not only technology or capital but add-on technology, distribution and manufacturing. This form of financing will continue to be emphasized in this country and elsewhere. There has been a tremendous amount of interest in the European Community as well as from Japan in gaining access to American technology when it is ready for the development stage under this strategic alliance/ corporate partnering format.

In summary, there are many sources of capital available today. Perhaps the challenge for us is to examine the issue of capital investment for technology development more from a private sector point of view as opposed to looking for the government to fund such initiatives.

BACKGROUND REFERENCES

Office of Technology Assessment. *New Developments in Biotechnology 5: Patenting Life.* Washington DC: OTA, 1989.

Organization for Economic Cooperation and Development. *Biotechnology: Economic and Wider Impacts.* Paris, France: OECD, 1989.

CHAPTER 23

The Future of the Food Industry in Mexico

Eneko de Belausteguigoitia

M exico is a land of contrast, both young and old at the same time. There is snow in the mountains all year round, there are tropical regions, desert areas where it never rains, as well as areas of tropical forest where it rains every day. The same is true of agriculture; although we still use the 8,000-year-old Egyptian plow, we have the latest technologies from the University of California at Davis.

Agriculture in Mexico is closely tied to history. Before the 1910 revolution, the main structure of agriculture was the "Hacienda." Haciendas were economic and political centers and were very productive. At that time the country's population was 10 million. More than 2 million people died during the revolution, which lasted from 1910 to around 1927. In 1930 we had the land reform, in which big properties were divided among workers. By this system, the "Ejido," peasants work a parcel of land that is government-owned. This has created social stability, but has made production very inefficient.

Private farmers are restricted to 200 irrigated acres or an equivalent of dry land. Most of the agricultural land is in the northwestern part of the country, which counts for about 80% of the production. Population has risen from 10 million in 1910 to 85 million this year, and this increase in population has created a great problem of food production. During the 1940s, 1950s, and 1960s we experienced growth with stability, and we were growing at a rate of more than 6% a year. In the next decade, however, the oil crisis occurred and the government began nationalizing everything. In the early 1980s about 70% of the Mexican economy was government-controlled, a much larger percentage than in some Communist countries.

In the last 6 years the government has begun to swing to the other side. Fighting inflation has become government policy and inflation has gone from 150% to approximately 20%. Another goal is the privatization of most industry except in the energy sector, which is to be kept in government hands.

Another important goal of the present administration is to generate steady growth. This has been impossible in past years because of our national debt and inflation. Mexico is now opening its borders and opening its economy by permitting foreign investment. This will help satisfy one of our basic needs which is creating jobs. We must create about 1 million jobs per year: this is quickly said, but it requires a great deal of capital and resources.

Another main problem is education. We must create places for about 2 million students per year in order to improve education. In the near future our commerce will have strong ties with Canada and the United States. We can clearly see what is going on in the Pacific, and thus we hope to increase our commerce by reducing our barriers.

The process of privatization of the sugar industry has begun. This year we acquired four factories that used to be state-owned. Our strategy for this year is to bring the four factories to full capacity. The second stage is to use subproducts of sugar cane to produce soybeans along with cane in order to optimize our resources. Thus, we can not only feed our cattle more efficiently, but also increase our meat production. We must provide more cash flow to the families of the workers: a proposal has been made to give goats to the wives of the cane growers so that they may produce cheese and a Mexican sweet called "cajeta."

Another goal is to give added value to the products we produce. We also must enhance energy production and control pollution. By using alcohol in automobile engines, we can decrease pollution, especially in Mexico City.

In the personnel area we want our workers and managers to improve the quality of their work. We believe that quality is the basis of growth. We also want to enhance the family, social and spiritual lives of the workers.

In the education sector we must train managers. Their effect in the total economy and in education within the country can be great. We now have a business school with branches in Monterey, Guadalajara, and Mexico City.

Mexico needs and welcomes increased business ventures, and better relations between our two countries will be to our ultimate benefit.

CHAPTER 24

The Food Industry in France and in The European Community

Francis Gautier

T rends in consumer habits are changing in France. People eat differently today from yesterday. In our language we call it the "destructurization of the meals." Family meals are replaced more and more by individual meals, by outside meals. Such changes influence our industries. The population is aging. In some countries, very rapidly. We have to follow it closely to offer products that fit consumer needs.

In France we noticed a very large part of baby food was consumed by the elderly. Therefore, we tried to market products directly oriented to them. It failed because the elderly do not want to acknowledge that they buy specific products because they no longer have teeth or they have stomach problems. They prefer to buy baby foods instead of buying specific products designed for them. It is probably a problem of communication. We have to adapt, and we are looking for ways to solve these kinds of marketing problems.

Biotechnology is an important and essential component of our industry. We use the fermentation process in beer, yogurt, cheese, and vinegar. In one year our manufacturing process will benefit from transformed microorganisms that will permit the acceleration of the manufacturing process or permit operation at lower temperatures than before. This biotechnological adventure will result either in savings in cost or in improved qualities, and both are of interest in our competitive businesses.

The revision of food regulations creates major problems in Europe. Most food regulations and laws date back to the beginning of the century. For example, in France the main law dates from 1905. The purpose of the law at that time was to require that products manufactured in factories be as similar as possible to those that our mothers manufactured in their kitchens. These laws were very static. They were not oriented to innovation and improvement.

New regulations have been established in Brussels. Every nation in the EC is obliged to adapt their national rules to Brussels' rules, which is very difficult. The complications we face are enormous, but we must harmonize the regulations. For example, we have to standardize the systems of control. In some European countries the quality controls of the authorities are made at the manufacturing level, while across the border the controls are established at the retail stage. We have not yet solved all problems, but about half of the 300 new regulations that have to be signed by the end of 1992 have already been signed.

Because of the change in regulations manufacturers will have more responsibility than in the past. For example, in France we are rather bureaucratic. We have tight control by the authorities. The manufacturer was safe as long as he complied with the rules and regulations. Tomorrow, with all the innovations that we expect, with all the new procedures that will be used, the government will not be in a position with its present staff to control everything in detail. Therefore, manufacturers will have much more responsibility.

The problem of labeling, which is very important in the United States, is probably of still greater importance in Europe. Consider the same type of products that are made in different countries. The contents are described by different units. Yogurt in the United States does not have the same definition as the yogurt in other countries. To ensure fair competition tomorrow, definitions must be made to ensure that consumers know what they buy, wherever it is produced.

If household garbage disposal is a serious problem in the United States where landfill is less and less available, it is the same in Europe. Food packaging accounts for a large part of household garbage and creates a risk to our industries. For example, if a country adopts a decision against the one-way bottle and beer manufacturers are forced to use exclusively returnable bottles, it gives an advantage to the local manufacturer and it prohibits competition from outside, which is completely contrary to Brussels law. We must find a reasonable solution to this problem.

The product circulation rule among the 12 countries in the common market is that if a product is legally authorized by one of them, the other countries cannot forbid the entry of that product into their own countries. Brussels has found that this solution accelerates the unification of the markets, because previously it appeared to be impossible to come to agreements on what they call vertical regulations (a common decision concerning every possible product). This new rule permits a product to move from one country to another as long as it is authorized in the country of origin. This rule is in effect and is efficient. Its adoption did not wait until 1992.

Our problems are really similar to problems in the United States. They are, nevertheless, a little more difficult because of the heterogeneity of Europe. Our continent consists of nations having their own structure and history and different food habits. Consider the way we eat cheese in Europe. It varies in every country. We do not eat cheese at the same time in the meal. Some eat at the beginning, some at the end. We do not eat the same type of cheese. Some eat cheese with forks; others, with bread or with biscuits and butter.

Food preferences and eating habits may never become really standardized. What we will expect to see and what we are preparing ourselves to live with are two different types of products. There will be national products in every country that fit the local tastes and historical culture. Simultaneously, some international brands will succeed more easily when they introduce new concepts rather than a revision of an already existing product.

For example, when Coca-Cola came with the U.S. Army after the war, there was no cola-based soft drink in Europe. Coca-Cola was then able to introduce the same product, the same packaging, the same concept in all European countries. It was adopted immediately with communication and advertising that was really similar everywhere. This is possible.

The same was true for the chocolate bar, which was a new concept. Chocolate bars were introduced in Europe with the same advertising communication everywhere. But for products such as beer or yogurt, the same products have neither the same positioning nor the same image in various countries, even if it is the same brand. This type of product has to be handled with care because it is very difficult to change a product image. These are the kinds of product difficulties faced by European businesses and are perhaps not as common in the United States.

When food product regulations are the same in the United States and Europe, we all will realize the advantage. We are forced to add specific labels when we export to the United States, or the reverse for you when you want to sell in Europe. Money will be saved through standardization, but standardization must be accomplished in the EC before it occurs internationally.

PART VI

ECONOMIC PREDICTION

ABSTRACT

Two Nobel Laureates, economist Kenneth Arrow and physicist Philip Anderson, have been working together under the auspices of the Santa Fe Institute since 1987 on a new approach to economic forecasting. By applying some of the theoretical tools of physical science to economics, they hope to help economists recognize complex and unexpected patterns in their data. Both physicists and economists examine complex systems with many parts that do not interact in a linear fashion. A fundamental difference is that economists deal with the actions of people and physicists with nonliving units. Furthermore, physicists invariably have more data and can check their theories against experimental evidence; economists operate from scant data and tend to spell out their theories and assumptions more explicitly. Arrow has always dealt with hypotheses of so-called rational expectations; such expectations may not lead to good economic predictions. But, since rationality is imperfect, how can one deal with it? Arrow and Anderson each provide their insights into the intrinsic problems of economic prediction, a major factor limiting business planning.

CHAPTER 25

Economic Forecasting

Kenneth Arrow

The uncertainty of the future is inescapable, one must think about it and arrive at plans for action. A statement attributed to a number of thinkers is, "Prediction is very difficult, especially of the future." Postdiction, knowing what went on in the past, is also difficult. The past, however, is our basis for understanding the future.

Economic forecasting is slightly less than an exact science. To predict next year's interest rates, prices, even gross national product, is difficult, although we are slightly better at the last. To say that we are uncertain about these issues is an understatement. We know very little about the relatively near future in those highly specific terms.

I became reasonably encouraged about economics during World War II. I left my graduate study to become a weather forecaster and learned that the natural sciences had no great built-in advantage over economics when it came to forecasting.

Among other things, I worked on statistical verification of weather forecasts. There were opportunities to do it in a highly controlled manner, and they showed that the best forecasters—not the average—could do better than chance only up to about 72 hours.

Since that time, elaborate numerical weather forecasting models have been developed based on deep, underlying physical principles. A lot of money has been spent both here and in Europe, and I believe that 72 hours now can be raised to 96 hours as a result of all the great improvement.

When we discuss forecasting, of most concern to the agricultural industries is the weather in the next season. Predictions on that scale are still pretty hopeless.

There are many reasons for the uncertainty about the future. Obviously one is that our knowledge of relevant laws is always incomplete. Another is an idea that penetrated Western thinking only in the 17th century, the idea of uncertainty, the idea of chance. This may be a modern survival of the old idea of the will of the gods, but, nevertheless, we have the idea of intrinsic unpredictability exemplified by tossing a coin.

A more recent development, which only goes back approximately 25 years, suggests that even if in some fundamental sense there are deterministic laws, and even relatively simple deterministic laws, it may nevertheless be true that the dynamics that come out of them are so complex that we cannot make good predictions. This has been argued in particular in connection with the weather, that the problems of weather prediction are far greater than lack of knowledge; in such a complex system they are intrinsic. Indeed one of the early papers on chaotic dynamics came out of a theoretical study of meteorology.

Much can be written about uncertainty, how to think about it, how to prepare for it, and how to react to it.

Two general statements among many can be made. One is almost banal is its simplicity and yet surprisingly hard to implement. The fact that the world is uncertain, that our plans are never going to have predictable consequences, must be recognized. We must face the fact clearly that uncertainty is inevitable. Great difficulty is posed by the idea that a forecast is uncertain. I think there must be something deep in the human psyche as it has evolved for whatever survival reasons that makes it difficult to grasp that uncertainty.

A forecast of future population is a number. Such a number has a band of uncertainty that will be recognized by any serious scientist. Any demographer will give a band, but the user tends to forget about it and to look only at the mean. I have served on several boards of directors and have heard forecasts of the companies' futures as numbers, not ranges. There are no statements of confidence about the numbers, only verbal commentary, which tends to disappear in your thought, even in the thought of someone who is accustomed to dealing with uncertainty.

One of the consequences of the fact that uncertainty is hard to grasp is that you frequently get a synthetic certainty, a projection of current hopes and fears into the future. If you read discussions of the future and of utopias, which are a very interesting kind of literature, you learn a great deal about the time at which the utopia was projected and very little about the time for which it was projected.

A projection that I remember was that of a very distinguished educator, President Hutchins of the University of Chicago. When he heard about atomic power, he suggested that people would have to work only an hour a week to satisfy all their needs, with the advent of such enormous energy, and that the main problem would become how to manage leisure.

In that particular case, a fairly simple set of calculations was all that was needed for a better forecast. Economists in Chicago showed, on the basis of the evidence available in 1950, that atomic energy was possibly going to be marginally cheaper than coal, but possibly not. Therefore, there was no possibility of a large gain to the economy from a widespread adoption of atomic energy. That was one of the better forecasts I must say.

A second remark of a rather different character is that in making projections, one must recognize that systems do have feedback tendencies, and even if we are not totally aware of them, catastrophic predictions are not likely to come true. Somewhere, someone will determine how to avoid events that are highly threatening. I do not want to overdramatize this issue. Feedbacks may be slow, may sometimes lead to destabilizing consequences, but some kind of feedback is likely. At one time, the Club of Rome predictions had a great deal of influence. I was startled by the influence they had because it was obvious many straightforward feedback mechanisms had been ignored. If the world was going to become highly polluted, actions would be taken to intervene, and if oil or other natural resources were going to become scarce, the price system, which was completely neglected in the Club of Rome report, would take effect. Resources would become more expensive, people would economize.

I have argued that the future is uncertain, which is not to say that we are completely without knowledge. We are likely to have some idea of broad trends, but my concern is that we are also likely to have some idea of the variability around those trends, and the variability is very important in our planning. What we can hope to know is the degree to which the future is likely to fluctuate and the patterns, to the extent that we can discern them, of that variability.

To know that there will be a recession next year would be valuable. Such a prediction is unlikely, but to know that we are likely for the foreseeable future to have an alternation of prosperities and recessions is also valuable knowledge for the planner, especially if you have some idea of the average recurrence time of these recessions and how large they are likely to be. In planning for business or other purposes, such knowledge is useful. Just to know that the weather in the northeastern part of the United States tends to alternate between cold and stormy periods and certain relatively benign intervals, without knowing when these periods will occur, is of great use in planning the heating systems in your home.

A more subtle prediction is the pattern of rainfall, in regard to water reserves, which have been cited as a possible major problem. If rainfall each year tends to be random, so that what happens one year does not predict what happens the next, over a period of time there is likely to be a certain amount of balancing. If, however, as some have proposed, wet years are followed by wet years and even more consequentially, dry years are followed by dry years, estimates of needed reservoir capacity would differ.

Knowing these patterns, even though they do not improve predictability at a given moment, may nevertheless be important in planning semipermanent structures.

As an analogy, let us consider the earthquake phenomenon. We know that earthquakes will occur. Predicting the time of an earthquake has proved to be an exercise in total futility, either in the short term or in the long term. Nevertheless, knowledge of the fact of the potential occurrence is of great use.

In particular, events at a given moment cannot usually be predicted. I am thinking particularly of the effects of innovation. For a number of years, economists have been aware of the contrast between two sets of observations. For example, we have experienced what appears to have been radically rapid changes in information technology. A state-of-the-art computer is obsolete in three years; systems change. Software capabilities are incredible, and the spatial compression of data seems unbelievable. All of these developments presumably would herald increased productivity.

Nevertheless, the last 15 years have been marked by a remarkably slow rate of growth in productivity, much less buoyancy in productivity than in the 15 preceding years. There appears to be a mismatch. My colleague, Paul David, has suggested that if one studies the period 1895–1900, a tremendous amount of literature will be found on the wonderful benefits that electricity will confer. Electricity was not only going to light houses, but it was also predicted to result in diffused power sources. A factory would have many small engines rather than one giant steam engine on the first floor with belting flying hazardously. A new world was predicted and indeed came into being. Most of the predictions were correct. The period from 1900 to 1910, however, showed extraordinarily low rates of growth in productivity as the dynamos and the electric networks were spreading.

The forecast was correct, but it was correct about the 1920s, not about the 1900s. I think this delay probably holds true for the information revolution. In time, we will understand the full implications of the information revolution and will apply the techniques in ways that we do not fully anticipate today.

A word about policies, planning, and decision-making in the context of uncertainty. Many actions are taken as a consequence of our inability to predict the future, no matter what the future might be. Holding a diversified portfolio of stocks is an obvious example. We would all invest in stocks with the highest rate of return if we were sure of the future, but even though we may think one form of security will outperform another, we know that there is enough randomness, enough uncertainty, so that we diversify.

Futures markets perform a similar function in the commodity business. Insurance in general serves this function. There would be no insurance if the future were certain. Either one would not buy insurance if one's future were known to be good, or conversely, if one's future were known to be bad, no one would sell it. Insurance hinges on the fact of uncertainty.

Unfortunately, the examples I have given are not particularly relevant for long-term uncertainties. These measures meet uncertainties in matters of months or years, but not much beyond that. The problem of meeting long-term uncertainty is mitigated by the fact that one's commitments are limited in time. Capital wears out and locations can be changed. These factors in our lives are subject to alterations with time. We are not confined to our current conditions, and that is our protection against the future.

As a general rule, the greater the uncertainty, the better to avoid large and irreversible commitments, to the extent that it is possible. When the famous 1930s gangster, Dutch Schultz, was dying, his incoherent last remarks were taken down by a stenographer. One of them was, "Don't make no bull moves." His words are a lesson for the kind of future that one might choose. Maintaining flexibility or keeping one's options open, is key in these matters.

Consider this example of avoiding commitments, a proposal that unfortunately has never been very popular. We have a flood problem in certain parts of the United States. One way to meet the problem is to build dams and try to control the flood from upstream. Another is to keep people out of areas that are likely to be flooded; that is, zone those areas against building. Evidence suggests that in many cases, this would be a far cheaper alternative. One means of meeting uncertainty is to stay out of the uncertain range.

Research—particularly basic research—is a means of maintaining flexibility. In research, one attempts to reduce commitment to irreversible investments and to increase commitment to opening new channels. There are losses connected with such policies.

I do not suggest that flexibility is a cure-all; some activities are subject to what the economist calls, increasing returns to scale. Some things are not well done on a small scale, for example, standardization of products. Every new industry faces the problem that standardization is itself a positive economic factor. At one time in our history, it was necessary to settle on a certain standard cycle number for alternating current, so that power grids could communicate with each other. We face the same dilemma today in computers until the architectures are standardized. There is a trade-off between inhibiting innovation, which relates to flexibility, and the immediate short-term benefits of standardization. There is no simple answer to the question. The typewriter keyboard, as some have determined, was standardized on what apparently is an inefficient system. Many people have tested it and we know that reorganization of keys on the keyboards can result in about 20% greater efficiency than use of the present configuration. But the advantage of training on a standardized keyboard, which can be used on a great many different machines, is obviously an advantage that has outweighed flexibility or receptiveness to new ideas or new configurations. I believe that if the difference were not 20%, but 60%, we would abandon the advantages of standardization.

Within this context, we can discuss the greenhouse effect. Although I am not a scientist, I have followed the literature. The facts about the greenhouse effect are uncertain. This is one of those rare examples in which the theoretical foundation has been known for almost a century, first advanced by a distinguished chemist, Arrhenius. On the other hand, the empirical evidence is difficult to interpret, and thus far, unambiguous evidence of warming is hard to find.

Nevertheless, one must reckon that this is one of the cases with high risks. If warming is going to occur, we are not sure how much, and it is not absolutely inconceivable that mean world temperature will rise by eight degrees centigrade in 75 years. Such a rise is very, very unlikely, but there are certain kinds of feedback arguments to support the warming. I do not suggest that is a realistic figure, but you will find serious people saying that it is possible. The idea, therefore, that we should take some cautionary steps, even now, is not a ridiculous one. On the other hand, there is a good argument for not taking drastic steps at high cost today against such an uncertainty.

Obviously, research is needed on the cycle of carbon dioxide usage. Research is also needed on the possible social and economic effects if there is a warming. Such effects are not very clear. The theory has been proposed that some of the effects are positive. I remember hearing a talk on this subject in the middle of a New England winter in which the speaker suggested the possibility of two degrees warming. Somebody looked out the window and said, well, are you sure that would be bad? There are positive economic values to a warmer climate. I don't think that very good research has been done so far on what either the agricultural or the urban effects would be if there were warming. There is need of research to prepare the way for substitute products in the event that certain kinds of agricultural production will be restricted. Finally, the capital intensity argument would suggest that cities that are in danger of flooding, for example, should be encouraged to proceed cautiously to improve port facilities, but not to grow and/or not to expose too much to risk.

BACKGROUND REFERENCES

Anderson PW, Arrow KJK, Pines D, eds. *The Economy as an Evolving Complex System.* Redwood City, CA Addison-Wesley, 1988.

Marschak J, Schurr S, eds. *Economic Aspects of Atomic Power.* Princeton NJ: Princeton University Press, 1956.

David PA. The dynamo and the computer: an historical perspective on the modern productivity paradox. *American Economic Review* 1990 May 80:355-361.

Diamond P, Rothschild M, eds. *Uncertainty in Economics: Readings and Exercises.* New York: Academic Press, 1978.

CHAPTER 26

Perception and Predictions

Philip Anderson

From time to time I have investigated the manner in which people think about the future and will first describe possible ways in which people might think about the future. There is or was a fashionable profession known as futurology. In the early 1960s, an entire issue of *Daedalus* called "Toward the Year 2000" was edited by Daniel Bell. Fifteen years or so later, there was an update of this issue.

Looking through my files I found that I lectured on futurology several times 15 years ago for a course on Science and Society at Cambridge University. I discovered that many of my thoughts about futurology then are germane today. For instance, in an assessment of the futures imagined by futurologists, one should identify not only the futurologists, but also those who support them. The Herman Kahns and Rand Corporations of the world are selling their product to great corporations, so naturally they extrapolate a massive growth of material production. Alternatively, there are skinny prophets living in the Third World. At the time I lectured on futurology, there was Ivan Illich, for example. Prophets such as Illich projected a future less and less dependent on knowledge and technology.

Technologists such as Dennis Gabor project marvelous technologic visions, and appear to imagine that technology will necessarily provide solutions to problems. They seem not to realize that technology is complex and interdependent, or that technology can be counterproductive as well as productive. For example, technology can produce problems such as crack, which is a very clever technologic breakthrough that opens an enormous new market for an old product, cocaine.

My lectures were presented at the time of the oil shock and at the time of the famous Forrester and Meadows theory of the Limits to Growth. This theory is in rather well-deserved disrepute today. At the time, a number of economists, in their arguments against this theory, suggested that the combination of technologic substitution and price mechanism would enormously mitigate the rate of our collision with the limits to growth.

At that time, at least six metals were projected to be exhausted by the year 1989. Among them were lead, mercury, zinc, gold, and silver. This prediction was, at most, somewhat exaggerated. In regard to minerals, we have substitution, the technology of exploration, recycling, the price mechanism, all of which combine to make mineral deficits among the least worrisome problems that we have.

Energy, I predict, is also amenable to the price mechanism. This proved to be true in the 1970s. We appear to be reverting to older ways, but again, as we find energy to be in short supply, I am sure we shall also find ways to substitute, replace, and explore.

Economists, however, have indicated quite correctly that for genuinely nonrenewable resources, a futurologist does not have to announce that a resource is about to be exhausted; one has only to look at the price. Economists have suggested that when actors in a given market see that the supply of the resource they trade is decreasing, the price will rise rapidly, well in advance of exhaustion. Thus we have an excellent means of predicting the exhaustion of a resource.

Twenty years ago it was less apparent that a great many amenity goods were indeed pricing themselves out of the reach of all but plutocrats: desirable real estate, for example, especially waterfront real estate, acceptable higher and secondary education, fine art, antiques, health care, genuinely fresher, tasty foods, and now we are seeing a fantastic escalation in the price of waste disposal.

A noteworthy and valid response from even the most conservative governments has been to reserve some of these scarce resources for the public at large, via parks and museums and development planning. This action has been taken in almost all advanced countries except the United States. Socialized medicine is provided in one or another guises, again with the United States as an exception. Other governments provide museums of various types, and scholarship programs for the public.

When we denigrate the Limits of Growth theory on the basis that the Malthusian catastrophe did not occur, we tend to ignore the fact that one after another of our resources is indeed diminishing; during our lives we have seen a rapid transition from general availability of the above kinds of resources to virtual unavailability except on a public basis.

Although each of these changes diminishes the quality of our lives a certain amount, it is still possible, out of season, to walk a deserted beach or to visit the less fashionable sectors of museums in reasonable comfort.

Food of course is a special case. Technology has served us well thus far and appears sufficiently advanced to accommodate enormous increases in the population of the world, if socioeconomic factors can be solved.

Futurologists inevitably underestimate the flexibility and pervasive effect of technologic change. No one predicted E-mail; no one predicted the Xerox, fiber optics, the many medical imaging technologies, the green revolution, the pill, RU-486, or the computer revolution. If one examines the technological predictions, even as late as 1960, you see few of the technologies that are most useful today.

No one can predict the impact of developments in progress, such as a dysentery vaccine, or RU-486, on the future of the developing world. Nor can we predict which problem may suddenly be alleviated or worsened by an unexpected technological development.

Present political or social trends can be extrapolated a few years into the future. Much of the future, for example, is present in the form of resources or population or technology, all of which are actually in the pipeline. You can see them, if you know where to look. For example, it was clear in 1973 that the world's energy supply was not going to collapse, and that it would ease in approximately 1980, which is what happened. It was also easy to see that if prices rose, great quantities of energy would be saved.

In spite of the creativity of futurologists, they appear always to underestimate the ability of the world, or great portions of it, to undergo revolutionary change. No one predicted such universal social phenomena as the student revolution that followed the great cultural revolution in China and seems to have been related in some way to it; the universal phenomenon of the Great Depression that had the lasting and rather revolutionary effect of leaving most of the governments of developed countries everywhere in effective charge of their economies. There was the Volcker–Reagan–Thatcher economic revolution that reversed inflation, to everyone's surprise, reversed the culture of public affluence, raised interest rates almost throughout the world, and changed real interest rates from negative, or neutral, to large and positive. The result is the disastrous situation in which our Third World friends now find themselves. Another revolution in progress that no one predicted, and of which we certainly cannot predict the outcome, is the Islamic revolution. It may be as destructive as any event in world history, or it may begin to die. The Islamic revolution seems to be associated with other fundamentalist movements. The apparently irreversible victory of the West in the Cold War of course is important, but is only one of many revolutionary changes that have occurred in the past decades and again was not predicted. These examples have been object lessons to demonstrate that history is not an inevitable progress in a straight line, but much more like a succession of avalanches.

One final footnote to futurology: although it is extremely difficult to predict the future, it also is surprisingly difficult to predict the past. We seem to know much less about the past than we normally think. Recent events in China indicate that we have been looking at China with blinders and that we knew little about the history of that country, about what was transpiring in one quarter of the world for 40 years. The flip-flops of media attitude toward China have been spectacular and probably none of our attitudes about China conditioned by the media has been correct. The media picture of the Soviet Union, of Iran, and of the United States has been equally questionable.

During the years that many of us regard as the glorious past of the United States, we were running banana republics, and the FBI was controlling a great many political events in this country about which we knew nothing. When our now-respected public servants were busily training the thugs who later formed Latin American death squads, we seemed to know nothing about these activities. It seems impossible to make sensible policy decisions except on a day-to-day basis until our knowledge base is greatly improved.

This then brings us to my second topic. Although the knowledge base available to our policy makers is weak, what I will call the concept base is even weaker. I will use economics as an example of a very general point of view that we have been trying to understand and through which we will attempt to predict the behavior of very complex systems.

Admittedly, the world economy or any national economy is a complex system, although not necessarily that much more complicated than evolutionary biology or the human immune system or neurobiology or even the biology of organismic development, which are all examples of subjects that we have studied recently in the Santa Fe Institute. We refer to these subjects as Complex Adaptive Systems.

The Santa Fe Institute program in global economies evolved from the belief that although economics is a discipline through which one may understand detailed situations or understand relatively short-term trends, it does not seem to enable understanding of the reason for the wild fluctuations that have characterized macroeconomics over the many decades of reasonably reliable history. Economics seemed to be concept-poor in the same sense I've described.

I will begin with a moral, which is complex in itself and consists of two parts. First, complex problems are not hopeless to study, and second, the nature of the answers derived from understanding may not be satisfactory at first and may not appear to be answers at all, but nonetheless may have real policy consequences.

I shall use two examples for my two morals. One I call scale-free behavior. The other is the consequence of what economists call increasing returns in conjunction with the complexity of the economic system.

I'll use earthquakes as an example. For many years, the law for the probability distribution of earthquakes as a function of the total energy in the earthquake has been known. It is called a power law and indicates that the probability of an earthquake that has a given amount of released energy is proportional to the inverse of the amount of that released energy. In other words, there would be one-tenth as many earthquakes with 10 times as much power, or one-hundredth as many earthquakes with 100 times the energy, and so on.

The law or theory sounds innocuous, but it predicts disastrous consequences. It says that the total energy released in earthquakes of a given size is the same in every decade of size. Thus, as much energy is released in earthquakes of size eight, which is 10 times as big as an earthquake of magnitude seven, such as the San Francisco earthquake; so the totality of energy in San Francisco-sized earthquakes is equal to the totality of the energy of earthquakes 10 times that size. That also means that the totality of energy in earthquakes 100 times as great is equal. We have never seen an earthquake 100 times as great but there is no sign in the distribution of earthquakes of any fall-off at some maximum size. Therefore, we have no reason to believe that there will not be an earthquake 100 times the San Francisco earthquake at some time.

The probability distribution of earthquakes is an example of scale-free behavior. It is defined as having no intrinsic scale, no visible maximum or minimum size, and no visible average size. It is a distribution of sizes, and looks the same in every scale that is used to measure it. Whether large earthquakes are described as six, seven, eight, or nine, is irrelevant; the distribution has little relationship to the scale by which it is measured.

Another set of beautiful examples of scale-free behavior is the so-called fractal that the mathematician Mandelbrodt has described. He has shown us that many landscapes, cloud shapes, and coastlines in the world are examples of fractal behavior.

Physicists are beginning to understand why fractals occur. Earthquakes, as a matter of fact, can be reasonably understood on a model of avalanche phenomena. Every earthquake represents a break in the earth's crust that propagates along the fault by avalanching, by one break weakening the segment next to it, which then weakens and breaks, which then weakens the next segment, and so on. Where it stops appears to be an absolutely random process.

A physicist, Per Bak, and his collaborators have shown that if you have a steady source of strain or energy, such as the gradual motions of the Earth, which can only really be released in such avalanches, it will inevitably lead to scale-free behavior of this kind. This was an enormously exciting new idea and its consequences are being studied throughout science.

Many data related to economics show scale-free behavior. Another well-known law is that for the sizes of cities, which is rather like that for earthquakes, that is, the number of people in cities over 10 million is equal to the number of people in cities between one million and 10 million is equal to the number of people in cities between 100,000 and one million and so on.

Another example of scale-free behavior which is less serious, but true, is that personal wealth in the advanced nations follows approximately a one over wealth to the third power law. The unpleasant aspect to the law is that it allows for the existence of excessive wealth in the presence of poverty, but it is not disastrous because no large fraction of the wealth is held by a few people.

On the other hand, the law for nations is much more severe, that is, a one over wealth. The top nations, therefore, monopolize a finite fraction of the resources of the world. I do not suggest that anyone knows the causes of such laws, but I do suggest that each has a rational explanation. The existence of such laws shows that economics is not yet a complete science.

Another law of this type, which I find fascinating, is the law of the random walk. This law states that the economies of all advanced nations fluctuate independently and randomly in a very specific way, called the random walk. This law is frustrating but also enlightening and useful. On a time scale through a few years, any one of our nations is precisely as likely to grow as to experience a recession.

The law of the random walk is completely independent of past history, as far as can be determined. It does not compensate for past defects. You cannot predict the sign of the fluctuations that are going to happen. In addition, it has the unpleasant effect that any fluctuation caused or experienced at any given time persists forever. For example, we have not yet regained the growth lost during the Great Depression. Therefore, an innocuous recession is growth lost forever.

Finally, we have the prevalence in economic systems of the phenomenon of increasing returns, that is, situations in which an increase in production makes it easier and cheaper to produce more, rather than the usual supply and demand theory, in which an increase in production causes an increase in price. We have tried to meld the phenomenon of increasing returns with another fact of economics called the economic or technological web, in which an increase in the supply of one good, call it A, may create a market for or make it possible to produce a second good, B; the existence of B makes another good possible, and finally good C feeds back on good A, and so on.

As an example, auto production required highways and made possible gasoline stations, fast food emporiums, and shopping centers, which added to the market for cars and for highways. The study of this networklike character of technological economic systems may well lead us to avalanching behavior of the sort that may lead to scale-free distributions and to random fluctuations in the economy, which, I repeat, has no satisfactory source in present-day economic theory.

To restate my morals, "The nature of the answers one derives in studying complex systems may not be useful in predicting behavior of the systems in the long run (although some parts of what we do are aimed perhaps at more sophisticated and subtle methods of predicting the course of the economy). Our predictions, however, are more likely to come in the form of statements about probability distributions, or the models which lead to appropriate probability distributions." With luck, we can prescribe the means by which such distributions can be changed and their extremes can be modulated.

The second moral is even more important, that is, "almost no system is too complex for its study to be rewarding, and most of the systems encountered in daily life are far too complex for the simplistic kind of straightforward extrapolation that characterized the old futurology systems."

We must expect enormous upheavals, qualitative changes, totally unexpected disasters, but also totally unexpected positive breakthroughs.

BACKGROUND REFERENCES

Anderson PW, Arrow KJ, Pines D, eds. *The Economy as an Evolving Complex System.* Redwood City, CA: Addison-Wesley, 1988.

[Anonymous]. Blueprint for survival. *The Ecologist.* 1972 January.

[Anonymous]. Doomwatchers and cheermongers. *Cambridge Review.* 1973 February 2.

Bell DI. *Toward the Year 2000.* Boston: Beacon Press, 1969.

Forrester TW. *World Dynamics.* New York: Wright-Allen Press, 1971.

Gabor D. *Innovations: Scientific, Technological, Social.* New York: Oxford University Press, 1970.

Illich I. *Tools for Conviviality.* New York: Calder Press, 1973.

Meadows DH, Meadows DL, Randers J, Behrens WW. *Limits to Growth: a Report for the Club of Rome's Project on the Predicament of Mankind.* London: Earth Island Ltd., 1972.

Conclusions: The Impact of New Technologies and the Function of the Global Food System in the 21st Century

Ray A. Goldberg

This paper will summarize the Conference participants' perspectives of the major economic trends, technological breakthroughs, and social and political changes that will affect the future structure of the global food system in the 21st century. To assess the potential changes in the global food system, one must have a reference point against which to measure potential changes in the structure, linkages, functions and kinds of participants in the food system. Table 1 shows the dollar changes in the major sectors of the food system over the last 38 years with projections for the next 40 years. By the year 2028, global farm supplies will have increased 16-fold from $44 billion in 1950 to an estimated $700 billion; farming will have increased 11-fold from $125 billion to $1,465 billion; and food processing and distribution will have increased 32-fold from $250 billion in 1950 to an estimated $8,000 billion. This half trillion dollar agribusiness economy of 1950 will become a $10 trillion economy by the year 2028.

What trends and private and public policies have helped shape the food system of the past and what structure will emerge in the future? Historically, from the end of World War II to the early 1970s, global grain surpluses in the United States acted as a shock absorber for the food system of the world. In many cases government programs determined price signals and most food processors and retailers really did not believe that they were part of, or needed to be part of, a vertically value-added food chain from input supplier to ultimate consumer. Raw materials were freely available. The 1970s changed all of that and suddenly one had to be aware of the fact that the shock absorbers were no longer present. Processors and retailers became aware that a shift had occurred to a seller's market and, thus, the quantity and quality of agricultural commodity products were no longer always assured. Food and energy prices fueled global inflation. The recognition in both developed and developing nations of the importance of a global agribusiness that utilized half the world's assets, half the world's labor force, and provided half the world's consumer expenditures, made that business a political and economic priority for all nations. Food security, nutrition, resource management, environment, food safety, efficiency, yield, and value added became important factors in evaluating food policy and the private and public policy changes of the food system.

The 1980s have led to volatile changes in supplies, with gluts and shortages occurring throughout the decade. In the midst of this new awareness of a global food system, scientific breakthroughs in molecular biology provided new knowledge of the component parts of plants, animals, and humans that enabled scientists to change the production process, nutritional components, and diagnostic procedures that improved the efficiency, quality, healthfulness, environmental aspects, and nutrition of the world's food supply. It also enabled land to become not just a food, feed, and fuel factory, but a potential pharmaceutical factory as well. In essence, this new scientific tool for the first time enabled farmers to adjust to changes in climate and to provide alternative ways to improve traditional crops and animals while creating nontraditional products in a more environmentally sound manner. This freedom and flexibility at both the farm and manufacturing level leads to the potential of tailor-made foods, feeds, fuels, and pharmaceuticals produced on and off the land. It also provides means for closer collaboration of all the segments of agribusiness, that is, farm supplier, farmer, food processor, and distributors.

One overwhelming conclusion that arises is that biotechnology is the most important scientific tool to affect the food economy in the history of mankind. Furthermore, these scientific breakthroughs are occurring at the same time other major political, social, economic, environmental and informational changes are also occurring. They include the following:

1. Major political changes in Eastern Europe that encourage private and public managers to cooperate in the development of the new scientific breakthroughs to help their agribusiness systems leapfrog into the 21st century. Similar political changes are occurring in the developing world, especially in Asia, with greater emphasis on the private sector and science to improve their food system. The early adoption of BST by both the Soviet Union

and the creative farm cooperative in India (Amul Dairy) are but two such examples.

2. These political changes have also led to freeing up of foreign exchange rates, interest rates, and commodity prices that have led to a more volatile economic environment. When this volatility is added to the unknown impact of major plant, animal, and bacteria innovations, risk management becomes a high priority for both public and private managers.

3. The new uses of the science of molecular biology are occurring at a time of budgeting restraints, and measurements of human and economic costs for health and nutritional benefit provide a new dimension to cost-benefit analysis that goes beyond traditional operating statements, shadow prices. Such costs become a major part of a social cost-benefit analysis.

4. Changing demographics in the developing world where population, although slowing, is expanding faster than the food supply and adversely affecting national and human resources needed to produce the food. These changes have led to demands for and challenges to new technologies that can improve both productivity and nutrition, and simultaneously attack environmentally unsound production and manufacturing processes. Another challenge for biotechnology.

5. The lack of understanding of this scientific tool by the media confuses them and the public as to the benefits and dangers of the new revolution. This confusion is compounded when both the private food system and the scientific community are timid in providing leadership in explaining the value of this new tool. The fears of scientists and company executives of accusations of conflict of interest inadvertently frighten the public as to the safety of the food supply and their particular products. The media exploit the consumers' fears and attempt to locate real or imaginary conflicts of interest for both the scientist and the food executive.

6. Although there has been a history of impartial bodies such as the National Institutes of Health, the Food and Drug Administration, or transnational scientific committees, that have conducted studies on new ingredients, new formations of foods, or new uses of proteins, genes, bacteria, or hormones, the credibility of such groups has been questioned by "public interest" groups who doubt the impartiality of such institutions. The question as to what institutions still have credibility in the United States and global food system is a difficult one to answer. The need to create new institutions was mentioned throughout the conference proceedings. Scientists also noted that they were reluctant and ill-suited to be the defenders of change or the defenders of the new technology.

7. In spite of the negative environment noted above, technology is moving forward more rapidly than the scientific community anticipated. Genetically improved plants will soon provide sweet grapefruit, improved wheats, nutri-

tionally balanced soybeans, and biodegradable plastic bags. In 4 or 5 years, genes will be replaced with better genes, using plants and animals as pharmaceutical factories for hormones, enzymes, and biological agents; all will become commonplace before the turn of the century.

8. Another trend is the need for new alliances in the new biotech food structure that is beginning to emerge. There are also concerns with respect to the perception of safety of these biotechnologically improved foods.

9. Resource management and the role of biotechnology in maintaining and improving water, land, and climate is another important priority.

10. Who owns biological resources—is it a nation? An individual? An institution? Or the creative scientist who discovers the existence of, or new use of, an existing gene, bacteria, protein, hormone, and so forth? What institution decides patent rights within and between countries?

11. Who is concerned with the displacement of people as the new technology is created?

12. How does one measure whether a new technology is size neutral?

13. How does one promote products that maintain and improve the health of the plant, animal, or human being—in essence preventive medicine—and respond to consumers' demands to stay well?

14. Who will provide the risk capital for basic and applied research that may not add to the bottom line for many years?

15. Who will train the new leaders and managers of private, cooperative, and governmental institutions who must understand both the new technology and its practical application?

16. Finally, who will be the "point" people to encourage the development and use of this technology? Do we provide only what the consumer thinks he or she wants?

Before attempting to assess the above trends, issues, and questions, let us first consider a number of major changes that affect the food industry and have occurred at the same time as the biotechnological revolution. These changes are five in number—branding, packaging, structural, demographic and informational.

Branding

Biotechnology enables us to produce noncaloric beverages and low-fat/low-cholesterol foods without adversely affecting the taste and texture of traditional foods. These foods can be tailor-made to meet particular nutritional, taste, and safety specifications. Some seed companies have already been approached to develop a "seed brand" for a specific type of cereal in cooperation with major cereal companies. These seed gene innovations are patentable under the plant variety protection laws. In this case, farmers are able not only to share in this increased yield of their crop, but also to share in the increased value-added nature of their crop. For the first time in history, farmers can brand their product and move away from the traditional commodity structure of their farm. Even the by-products of grain at the farm level, such as corncobs and cornstarch, enable environmentally sound products, such as biodegradable plastic bags, to be produced, which, in turn, provide additional outlets for farm production. Again, the by-products can be branded. Similarly, unique feeds have been produced that, when fed to chickens, lower the cholesterol content of eggs or help produce eggs that when consumed, help lower the cholesterol of the consumer. These new feeds exist in Europe and produce eggs that sell for twice the price of regular eggs. Similarly, nonresidue grapes sell for 20 cents per pound more in one supermarket chain. Not only are farmers producing food, feed, and fuel, but pharmaceutical products as well. The farm is fast becoming a plant factory that is capable of producing brandable differentiated products and by-products. In addition, in animal agriculture, growth proteins not only increase livestock, poultry, and fish yields, they do so in a way that increases the meat content and lowers the cholesterol levels of these foods.

Historically, the farm commodity surpluses of the past tended to rely on commodity-processed, private-label products to help dispose of unwanted surpluses. Most of the time, these surpluses were distributed through farm cooperatives who were more interested in providing a "home" for their farmer members' products than in developing a market. Today, the farm community wishes to improve both yield and quality, and to add unique attributes to farm products. The farm community has witnessed the success of the Ocean Spray Cooperative approach and wants to use that type of product differentiation as their model.

The ability to control quality and value from farm inputs to final consumer products is creating new joint ventures between manufacturers, input suppliers, and farmers. Additionally, new joint ventures between retailers and farmer-owned manufacturing operations are occurring. In California alone, five regional chains own a common dairy and a common orange juice operation, which, in turn, has contractual relationships with the farmers. These relationships also exist overseas. One supermarket chain provides orange trees and production boxes to the Chinese, who, in turn, provide labor and land to produce mandarin oranges of a certain quality for that chain. Similarly, Eli Lilly & Company and Amul Dairy of India have recently signed a memorandum of understanding for the development and registration of a range of biotechnologic products to be used for dairy cattle and buffalo in India, including BST and other new agricultural products of biotechnology.

The company also has made approved sales of BST to the USSR, Czechoslovakia, and Mexico with excellent productivity and quality results in all four countries. The NutraSweet Company of Monsanto, having created a branded sweetener, is working on other branded new ingredients and new foods for the 21st century. In essence, the technological, legal, patentable nature of tailor-made seeds, animals, and major ingredients in the global food system have added a new and increasingly important brand partner in the value-added food chain. The willingness of the consumer to pay a premium for these quality, nutritional, and safety-oriented products is, of course, key to the ultimate funding of these new joint ventures and these new companies. (See Table 2 for typical profit margins.)

Packaging

A second major change that is occurring is in packaging. The aseptic package has had a major impact on the shelf life and quality of other products. For example, in New York City at the present time, Minute Maid is test marketing a plastic aseptic package for their premium orange juice. I am told that on a test basis, they sampled the product after 120 days of storage and found the quality identical to the highest quality orange juice being put into the package at the start of the packaging process.

There are firms, such as the Don Watt Company, whose case we use in our agribusiness course, that specialize in changing either the label or the package or both for a product. They can enable that product to move from a number two position in the world not by changing the product, but by changing the label and/or the package. These types of consulting firms work with both store-owned brand or manufacturer's brand operators. In essence, the package and the label have become as important as the product in communicating to consumers a genuine response to their taste, nutritional, health, and safety priorities. In addition, some stores have created a whole new line of products responsive to consumer concerns such as Loblaw's President's Choice, as well as new Loblaw's President's Choice Green which are environmentally friendly products that include safer, natural, nontoxic, nonresidue pesticides, organic vegetables and biodegradable plastic bags and containers.

Structure

New structural changes are being fueled by new global competition, consolidation, volatilities, and a new market orientation of marketing boards, industry associations and private–public cooperative national food policy institutions. In 1988 alone, acquisitions of United States' companies by foreign investors totaled $33 billion by the United Kingdom (such as Grand

Metropolitan's acquisition of Pillsbury and Christian Brothers), $10 billion by Canada, $8 billion by Japan, $7 billion by France, and $4 billion by Australia. In the value-added agribusiness system, 25% of the farm inputs are produced by non-United States firms. Some 1% of United States farmland and some 20% of both United States food manufacturing and retail distribution is owned and operated by non-United States firms (See Table 3).

Consolidation is occurring at every level of the global value-added food chain from input suppliers to distributors. Even at the farm level, 133 million of the 155 million farms in the world have less than 10 acres of land and 1 million of them produce most of the commercial farm sales. In the United States alone, one out of two farmers receives 85% of his or her income from nonfarm sources. The top 200,000 super-family farms produce most of our farm products. Not only are the functions in the food system being performed by fewer decision-makers, but the logistic system is being shortened; some of the traditional distributors are being eliminated and replaced by more direct sales, contracts, and/or even joint ventures.

While these consolidations are occurring, the price volatilities in the basic commodities of the food system are increasing because of the dramatic shifts in grain inventories, and a reduction of 180 million tons globally in one year (130 million tons in the United States alone) as well as the freeing-up of foreign exchange rates, interest rates, and the lowering of the level of government programs. These price volatilities have all added new uncertainties in the food system. New risk management tools are being created to manage the new uncertainties together with futures markets that operate 7 days a week, 24 hours a day, to provide ways to minimize the risks. These risk management tools are used by both the buyer and seller and thus, in many cases, become another way for the buyer and seller to cooperate in their trading functions.

Not only is consolidation occurring on a vertical individual-firm basis, but horizontal consolidation is occurring also. For example, it was recently announced that a joint venture of British, Dutch, Italian, Swiss, and German supermarkets has taken place, providing one group with multibillions of dollars of purchasing power. In addition, vertical arrangements are being encouraged by firms that specialize in bringing buyers and sellers together in joint product promotion operations, even joint research and development of products. The learning curve advantage of these new experimental horizontal and vertical arrangements will permit the firms to view each other as partners rather than adversaries in trying to meet the consumer's new priorities in the global food system.

Many of these firms have their origin as quasigovernmental bodies who have become market-oriented to the point of creating their own global brands such as the Dairy Cooperative and Marketing Board of New Zealand. Other countries are trying to develop a national awareness of their food products, such as the British food processors and the government of the United Kingdom. Still other governments have created industrial parks in their country to encourage global scientific cooperation in biotechnology such as the Biotechnology Park in Osaka, Japan.

All of these new alliances indicate structural changes that encourage more direct relationships between the participants in the vertical value-added food system (see "technology leaders," Table 4) and their public and farm cooperative potential partners.

Demographics

The most important trend that drives the other revolutions is the change in global and domestic demographics. Globally, the engine for change is Southeast Asia. Currently, Southeast Asia has 30% of the world income. Over the next 40 years, it will have 60% of the world income. Major southeast Asian countries are Korea, Malaysia, Thailand, The Philippines, India, and China. This part of the world will be the engine of change affecting both our commodity and value-added product markets.

Information

Domestically, consumer demand for quality, convenience, nutrition, health, and safety, in addition to products that are environmentally sound, has led to minimarkets that are housed within superstores. Minimarkets respond to these various market segment demands. The information revolution and the ability to know instantaneously what each consumer wants as well as when and how these wants can be translated into demand have led to an increase in the power of the retailer, who plays a more important role in creating brands that not only reflect the image of the store, but also respond directly to the market segments the store wants to target.

HOW DO FIRMS RESPOND TO THESE REVOLUTIONS?

I have selected six different examples of how individual firms or collections of firms are looking at these changing phenomena:

Marks & Spencer and Loblaws

Their approach is to be the leader of change in the value-added food chain, and to create unique brands, such as St. Michael's, President's Choice, and Nature's Choice, that can provide premium brand leadership for the market for bodily and environmentally friendly products.

Unilever

This firm has acquired a major biotechnology research plant in the United Kingdom through which to research and develop tailor-made foods for its total organization.

Pioneer Hi-Bred Seed Company

This company is revising its research and development plans to help the farmer not only increase yield, but also quality, and with joint ventures in mind, to tailor-make foods for either the manufacturer or the retailer or both.

Conagra

This is the largest farm supply company in the United States, with over $1 billion in sales at the farm level alone. It is also the second largest poultry and the largest beef processor and flour miller in the United States, providing added value at each level of the vertical food chain. It is global in nature and works closely with the retailer, as well as providing a leadership role in market intelligence.

Grand Metropolitan

Their publicly stated mission is to be a global, low-cost, brand-leading firm that develops new products, and they have done so by revitalizing the Pillsbury Company and other U.S. food and drink acquisitions.

Consortiums

Finally, consortiums, such as noted in the *Wall Street Journal of Europe*, on May 19–20, 1989, are emerging. The article, titled "Argyll Studies Links with Ahold, Casino," states:

> London—Argyll Group PLC said it signed accords with two European food companies that may lead to cooperation agreements. Argyll, which trades as Safeway Stores in Europe and the U. S., said it will examine commercial opportunities with Koninklijke Ahold NV of the Netherlands and Etablissements Economiques du Casino Guidhard Perrachon & Cie. of France. Links among the three concerns could be made in marketing, distribution, production, development and exploitation of store formulas, Argyll said. The companies will also consider ties in management information systems and other computer applications. Ahold, a retail grocer that owns the Albert Heijn supermarket chain, had sales of 15.3 billion guilders ($7 billion) last year. The company owns substantial food-processing capacity and is an institutional catering supplier. Casino is a food retailer that operates a variety of stores, from

convenience outlets to hypermarkets. It also processes food and has commercial restaurant operations. Casino had sales of 51.4 billion French francs ($7.8 billion) last year. In the U. S., Casino owns restaurants, retail stores and cash-and-carry warehouses. Ahold operates in the U. S. through Bi-Lo stores, Giant Food Stores and First National Supermarkets.

From this article, it is evident that these firms are already planning for a more unified Europe in 1991. They have already been joined by an Italian supermarket chain since the news release.

All of these strategies call for new working relationships with all segments of the vertical value-added food system as well as the ability to take each firm's strengths and determine how to use the products of biotechnology, information, packaging, and structural change to meet the needs of an ever-changing global market. Firms that obtain a competitive advantage in anticipating change and choose wisely the appropriate partners to respond to that change will be the future global leaders of the food system. Technology will provide real product differentiation and real cost advantages (see Table 5).

Before turning to the implications of these trends and changes on the future structure of global agribusiness, one must also note the disenchantment present when one discusses biotechnology.

DISENCHANTMENT WITH BIOTECHNOLOGY

A variety of factors has contributed to investor, corporate, and public disenchantment with biotechnology. The conventional wisdom suggested that biotechnology would be used more rapidly in agriculture than in medical sciences. This belief was based on the history of plant and animal technology patents versus human health patents. What escaped recognition, however, was that the consequences of genetic technology could invade the total environment, rather than just a specific plant or animal, whereas, medical biotechnology that affected human life could be confined to a single individual. Therefore, agricultural biotechnology had to be viewed differently to ensure that the public would be free from unforeseen releases of new bacterial plant and animal life.

Not only did governments have to create a new set of regulations for this technology, but they had to do so in a way that was acceptable both nationally and globally. Delays in field testing were common in the past, but recently trials have become relatively more routine.

The science itself, in many cases, exceeds the ability of government agencies to review it. In some cases, the new technology required more labor input by the farmer and the application had to be modified to make it more "user friendly." Some terminology such as "somatotropin" or "growth hormone" rather than "growth protein" conjured negative images in both the producers' and consumers' minds. At the same time, new ways of measuring movement of bacteria and trace chemical detection gave regulators a more efficient means to track the experiments implemented through biotechnology. Biotechnological advances, in part, were contributing to their own measurement.

The long development time and the corresponding increase in costs for investment have led both the venture capital community and public investors to discount stock values of companies which were primarily engaged in agribusiness biotechnology. It was and is difficult for them to value a tool that only becomes useful as part of the total improvement of the value-added food chain. It is difficult to measure who will gain the most in that value- added chain—the new entrants or the existing players. Thus far, the disillusionment is widespread and even the biotechnology research and development divisions of some companies, such as the Corn Products Corporation (CPC) and Campbell Soup have had their technologies donated to universities or spun out into new corporations such as DNAP.

Nevertheless, some companies believe strongly that the very existence and future of their companies is highly dependent on the useful application of biotechnology to the development of new processes and products for their companies.

IMPLICATIONS AND CONCLUSIONS

Having examined the trends, changes, and even some of the disenchantment that relates to the biotechnological revolution that has and will affect our global agribusiness system, how will this system function in the 21st century?

It is presumptuous for one individual to assume that he or she can predict the structure and functioning of a global food system that encompasses half the world's economy as we approach the 21st Century. I can only provide one person's understanding of the strategies, priorities, and action plans of individual private, cooperative and public leaders who through their actions and responses to global consumer needs will help shape the food system in the 21st century.

1. On the global front, plant and animal genetic resources will be better preserved through greater cooperation among private seed and animal breeders, and national and global institutions and foundations. An organization called the International Board of Plant Genetic Resources (IBPGR) has been organized, financed, and directed by the Consultative Group for International Agricultural Research (CGIAR), the coordinating body for the network of international agricultural research centers such as CIMMYT (International

Maize & Wheat Improvement Center), IRRI (International Rice Research Institute), and ICRISAT (International Crops Research Institute for Semi-Arid Tropics). IBPGR has acknowledged the importance of breeders' rights (patent protection) and that of farmers' rights (the preservation of basic germ plasm in both developing and developed countries). This organization is currently housed in FAO (the Food and Agriculture Organization of the United Nations). It probably should be set up as a separate entity of the United Nations to avoid political pressure groups that may exist within FAO. In addition to germ plasm banks, the development of new gene combinations will undoubtedly lead to gene banks and to breeders' and farmers' rights to the development of new gene combinations. The agricultural extension service established in the 1860s played an important role in the development of plant and animal genetics, and their applications over the last 130 years of U. S. agribusiness development and the duplication of their efforts in other developed and developing countries will have to be complemented by the creation of such a service in the biochemistry departments of universities where familiarity with the medical and pharmaceutical world is greater than with the agribusiness world. The development of a liaison individual in each department will soon be commonplace.

2. On the regulatory front, scientific, national and global boards will be created to help provide common agreement on safety, in terms of testing and utilizing new ingredients, new processes, and new combinations of genes, enzymes, hormones, proteins, and so on. Conflicts of interest will always occur and such board members must refrain from participating in decisions that affect their own research or applications of their research (see Table 6).

3. The global agribusiness estimates in Table 1 indicate a more rapid increase in the value-added segment of the food system by the years 2000 and 2028 because of the further enhancement of the food supply in terms of nutritional benefits and breadth of choice. At the same time, scientific breakthroughs will enable the farm supply sector to increase more rapidly by the years 2000 and 2028 and the nature of those supplies will have shifted from a chemical base to primarily a biotechnological base. Although the farming sector may not grow as fast in dollar terms because of increased cost savings, the diversity of products produced on the farm will increase as will the value-added nature of those products.

4. Because the source of change can now occur at any and every level of the vertical value-added food chain, alliances among the participants will be commonplace. Those firms that cannot only be global, brand-oriented, new product developers, and least-cost operators, but who also can provide alliance and joint venture leadership will be the new leaders of the global food system.

5. Finally, technology will change half the world's economy, affect the

nutrition and health of populations, and affect the economic development and linkages between and within agribusiness and the rest of the global economy. Thus, private firms will be asked to play a more statesmanlike role in helping to coordinate a global food system as it responds to the political, social, and economic needs of society. It will not be enough to produce what consumers think they want or to worry about the bottom line, quarter by quarter. Rather, it will be necessary to create new products that improve the health and well being of consumers and their environment and to do so in a way that enables the firm to be a responsible and profitable leader in a global food system. Similarly, a scientist who, in the past, felt no obligation to be the "point man" for new technology will be expected to be able to relate his or her technological development to the needs of society and to provide the educational perspective that will enable society to utilize that new technology without fear and distrust.

Table 1   Global Agribusiness Estimates (billions of dollars)

	1950	1960	1970	1980	2000	2028
Farm supplies	44	69	113	375	500	700
Farming	125	175	255	750	1115	1465
Processing and distribution	250	380	6000	2000	4000	8000

Source: Author's estimate based on discussions with USDA economists.

Table 2   Profit Margins for Selected Companies

Company	Business	Margins	Year
Pioneer	Hybrid seed	28.7%	1983–87
Monsanto	Agricultural chemicals	25.6%	1985–87
NutraSweet	Artificial sweetener	19.5%[a]	1985–87

[a] Cash flow exceeded 40% of revenues.

Table 3   U.S. Agribusiness Owned by Overseas Firms

Major Sector	Percentage
Farm supplies	25
Farming	1
Manufacturing	20
Retailing	20

Table 4   Technology Leaders

1. Boutiques	DNAP, CALGENE, BIOTECHNICA relationships with Campbell Soup, DuPont, Continental Grain, General Foods, Ciba-Geigy, Brown & Williamson Tobacco Company, Farms of Texas, Kallihamns A.B., and Eastman Kodak
2. Ag-Chemical Life Sciences	Monsanto, ICI, DuPont, American Cyanamid, Eli Lilly, Sandoz, Dow
3. Traditional Plant Technology	Pioneer HiBred
4. Tailor-Made Food Advocates	Unilever
5. Traditional Processors	ADM & Ferruzzi
6. Farm Coops	Saskatchewan Wheat Pool, Ocean Spray, Welch, Land O' Lakes, and Agway
7. Retailers	Marks & Spencer, Loblaws
8. Specialists	NutraSweet division of Monsanto

Table 5   Power Shifts of Brands in the Food System: Technology Changes in Each Part of the System

Retailer	Loblaws–President's Choice Green Products—unique environment and nutritional leadership
Manufacturer	Coca-Cola—new aseptic orange juice package
Farmer	New unique fruits and vegetables, joint ventures on dairy and orange juice products
Farm supply	Pioneer Hi-Bred—developing new, more nutritional and chemical-free seeds
Chemical & Pharmaceutical Company	Monsanto–NutraSweet—low-calorie and low-fat foods

Table 6   Selected Agencies Interested in Monitoring and Harmonizing Product Approvals

Food and Drug Administration (FDA)
Environmental Protection Agency (EPA)
Center for Disease Control (CDC)
U.S. Department of Agriculture (USDA)
Organization of Economic Cooperation and Development (OECD)
Inter-American Institute of Cooperation for Agriculture (IICA)
U. S. Patent Office

In all cases, these agencies treat biotechnology as an extension of an existing science and that the products, therefore, are primarily risk predictable.

Biographies of Contributors

PHILIP W. ANDERSON, Ph.D., Joseph Henry Professor of Physics, Princeton University. Dr. Anderson also was Professor at Cambridge University and was on the staff of the AT & T Bell Laboratories in the Physical Research Laboratory. His awards include the Nobel Prize and the National Medal of Science. He is a member of the National Academy of Science, the Royal Society (London), the Academia (Rome) and the Japan Academy.

KENNETH J. ARROW, Ph.D., Joan Kenney Professor of Economics and Professor of Operations Research, Stanford University. Dr. Arrow is also External Professor, Sante Fe Institute and Senior Fellow by Courtesy, Hoover Institution on War, Revolution and Peace. He held prior faculty positions at the University of Chicago, Harvard University, Center for Advanced Study in the Behavioral Sciences, Churchill College (Cambridge), The Institute for Advanced Studies (Vienna), Massachusetts Institute for Technology, and the European University Institute. He has been awarded the John Bates Clark Medal of the American Economic Association and the Nobel Memorial Prize in Economic Science.

DURWARD F. BATEMAN, Ph.D., Dean of the College of Agriculture and Life Sciences at North Carolina State University. He was previously on the faculty of Cornell University, where he was Chairman of the Department of Plant Pathology. Dr. Bateman is a Fellow of the American Phytopathological Society and is listed in Who's Who in America and Who's Who in Frontier Science and Technology.

ENEKO DE BELAUSTEGUIGOITIA, Ph.D., chairman of Alfa Elai S.A., and cofounder of the Instituto Panamericana de Alta Direccion De Empresa, Mexico's only business school. Dr. De Belausteguigoitia also is on the Board of Governors of Universidad Panamericana and is a member of the board of the Colegio de las Viszainas. His crisis management skills played a vital role after the 1985 earthquake in Mexico City, where he was instrumental in organizing relief activities. He owns and manages several agricultural businesses, including the Constancia sugar mill.

DANIEL BELL, Ph.D., Henry Ford II Professor of Social Sciences, Harvard University. Dr. Bell held previous faculty positions at Columbia University, the Center for Advanced Studies in Behavioral Sciences, Salzburg Seminar in American Studies, University of Chicago, Russell Sage Foundation, and Cambridge University. In addition to being a Fellow of the American Academy of Arts and Sciences, the American Philosophical Society, and the Sociological Research Association, he was cofounder of *The Public Interest*, and has served on the editorial board of *Daedalus* and *The American Scholar*. He is the recipient of six honorary degrees.

D. THEODORE BERGHORST, M.B.A., Founder, Chairman and Chief Executive Officer of Vector Securities International, Inc., an investment banking firm involved with companies in the biotechnology and health care sectors. Prior to forming Vector, he was a Managing Director, Kidder Peabody & Co., Inc., and also held senior financial positions with INA Corporation's Lawrence Systems and Citibank, N.A.

WINSTON J. BRILL, Ph.D., president of Winston J. Brill and Associates, consultants in research productivity and creativity. He also serves as Adjunct Professor, Department of Bacteriology, University of Wisconsin, Madison, where he held a chair until 1984. He was previously a founder and Vice-President of Research and Development of Agracetus. Dr. Brill is a member of the National Academy of Sciences and was featured as one of the top ten innovative scientists by Business Week.

GEORGE E. BROWN Jr., Member of Congress. Mr. Brown has represented the Riverside–San Bernardino–Ontario area of California since 1972. He is also a senior member of the Agriculture Committee, is chairman of the Subcommittee on Department Operations, Research and Foreign Agriculture, and is the ranking Democratic member of the House Science, Space and Technology Committee. Mr. Brown's primary legislative interests for many years have been science and technology policy.

SIR ROY DENMAN, Senior Fellow at the Kennedy School of Government at Harvard University, and principal of Denman & Partners, L.P., a Washington- and Brussels-based consulting firm. Sir Roy has held numerous diplomatic and trade-related positions in Great Britian and Europe, including Head of the Delegation of the Commission of the EC to the Unites States, Chief Negotiator for the EC in the Tokyo Round of Multilateral Trade Negotiations, and Deputy Secretary of the Department of Trade and Industry. He is a Knight Commander of the Order of the Bath and Companion of the Order of St. Michael and St. George.

JOHN T. DUNLOP, Ph.D., LaMont University Professor at Harvard University. Dr. Dunlop previously was Professor of Economics and Dean of the

Faculty of Arts and Sciences at Harvard University. He has served every President since Roosevelt, including appointment as Secretary of Labor under Ford. His honors include the Murray, Meany, Green Award of the AFL-CIO, and nomination to the National Housing Hall of Fame. He is a member of the American Academy of Arts and Sciences, the American Philosophical Society, and the National Academy of Arbitrators.

GERALD E. GAULL, M.D., Vice President for Nutritional Science, The NutraSweet Company, and Adjunct Professor, Northwestern University School of Medicine. He was previously on the faculties of Harvard University, Columbia University, and Mt. Sinai School of Medicine. His academic career spans 25 years in research and teaching with emphasis on nutrition. Dr. Gaull's honors include the Borden Award in nutrition and the Gold Medal of St. Ambrosiano of the City of Milan.

FRANCIS GAUTIER, Vice-Chairman of the Board of Directors of the BSN Group, France, and Chairman of its American subsidiary, The Dannon Company. Mr. Gautier previously spent 25 years with Colgate-Palmolive, becoming Chairman and CEO of the company's French and North African subsidiaries. He serves as Chairman of Entreprise et Progres, a French association of business managers, and as Chairman of the Confederation of the Food and Drink Industries of the EC. Mr. Gautier has been awarded Chevalier de la Legion d'Honneur.

DONALD A. GLASER, Ph.D., Professor of Physics and of Molecular and Cell Biology, University of California, Berkeley. Dr. Glaser is also Chairman of the Board of Scientific Advisors of the Cetus Corporation and is a member of the advisory and review panels for the Department of Education, the National Science Foundation, the National Institute of Health and the White House. He was a member of the IBM Corporation's Scientific Advisory Committee. Dr. Glaser received the Nobel Prize in Physics and the Golden Plate Award of the American Academy of Achievement.

RAY A. GOLDBERG, M.B.A., Ph.D., Moffett Professor of Agriculture and Business, Harvard University. Dr. Goldberg has extensive experience in developing case studies on all aspects of agribusiness. He serves as a director, trustee, or advisor to many business, governmental and non-profit organizations such as Pioneer Hi-Bred International, Arbor Acres, John Hancock Agricultural Committee, Rabobank, Transgenic Sciences Inc., Caribbean Basin Initiative, and Beth Israel Hospital. He is the first President of the International Agribusiness Management Association.

JULES HIRSCH, M.D., Sherman Fairchild Professor and Senior Physician, Rockefeller University and the University Hospital. Dr. Hirsch holds concurrent medical positions with Columbia University, Cornell University Medical College, and the New York Hospital. He received the Dististinguished Alumnus Award, Southwestern Medical School, and the McCollum Award of the American Society for Clinical Nutrition.

ARTHUR KORNBERG, M.D., Professor Emeritus (active), Department of Biochemistry, Stanford University. Dr. Kornberg served as Professor of Biochemistry since 1959 and was Chairman of the department from 1959 to 1969. He previously served as Professor and Head of the Department of Microbiology at Washington University. He received the Nobel Prize in 1959 for his research on the enzymatic synthesis of nucleic acids.

NORMAN KRETCHMER, M.D., Ph.D., Professor, Department of Nutritional Sciences, University of California, Berkeley, and Director, Koret Center for Human Nutrition, San Francisco General Hospital. He has served as Director, National Institute of Child Health and Human Development and as Chairman of the Department of Pediatrics, Stanford University. Dr. Kretchmer is a member of the National Academy of Medicine of the National Academy of Sciences, and is recipient of two honorary degrees. His awards in nutrition include the Mead-Johnson Award and the Borden Award.

PHILIP LEDER, M.D., John Emory Andrus Professor of Genetics and Chairman, Department of Genetics, Harvard Medical School, and Senior Investigator, Howard Hughes Medical Institute. Dr. Leder previously served as Chief, Laboratory of Molecular Genetics, in the National Institute of Child Health and Human Development. His awards include the Bristol-Myers Award for Distinguished Achievement in Cancer Research, The Giovanni Lorenzini Foundation Prize for Basic Biomedical Research, the V.D. Mattia Award of the Roche Institute of Molecular Biology, the Albert Lasker Medical Research Award, and the National Medal of Science. He is the recipient of three honorary degrees and is a Trustee for the Massachusetts General Hospital and the Rockefeller University.

MICHAEL A. MILES, President and Chief Executive Officer of Kraft General Foods. Mr. Miles previously held several other executive positions in the food industry, including President and Chief Operating Officer of Kraft, Chairman of Kentucky Fried Chicken Corporation, and he spent 10 years with Leo Burnett Advertising. Mr. Miles also serves as a Director of First Chicago Corporation and Capital Holding Corporation, and is on the business advisory council for Carnegie Mellon University.

SANFORD A. MILLER, Ph.D., Dean, Graduate School of Biomedical Sciences and Professor, Departments of Biochemistry and Medicine at the University of Texas Health Science Center at San Antonio. He previously served as Director of the FDA's Center for Food Safety and Applied Nutrition and as Professor of Nutritional Biochemistry at the Massachusetts Institute of Technology. Dr. Miller's awards include the Award of Merit of the Food and Drug Administration and the Conrad E. Elvejhem Award for Public Service of the American Institute of Nutrition.

JOHN A. MOORE, D.V.M., President and Chief Executive Officer of the Institute for Evaluating Health Risks. Prior to this, Dr. Moore was Assistant Administrator of the Office of Pesticides and Toxic Substances at the EPA. He is credited for defining EPA's policies in the area of biotechnology and the development of scientific policy for the agency's use of risk assessment. Dr. Moore also served in various management capacities in the area of toxicology of the National Institute for Environmental Health Sciences.

IRWIN H. ROSENBERG, M.D., Professor of Physiology, Medicine and Nutrition, Tufts University, and Director, USDA Human Nutrition Center on Aging, Tufts University. Dr. Rosenberg also has held faculty positions at Harvard Medical School, Weizmann Institute of Science, and the University of Chicago. He has served as Chairman of the Food and Nutrition Board of the National Academy of Sciences, and as President of the American Society for Clinical Nutrition, as well as on many other advisory and editorial boards. His awards include Grace Goldsmith Award of the American College of Nutrition and the Robert E. Herman Memorial Award of the American Society for Clinical Nutrition.

JUDITH S. STERN, Ph.D., Professor, Departments of Internal Medicine and Nutrition, and Director, Food Intake Laboratory, University of California, Davis. Dr. Stern has received a number of awards and has served on numerous advisory boards, including the Human Nutrition Research Advisory Board, USDA, and the National Academy of Science Committee on Nutrition Components of Food Labeling. In addition to her numerous research publications in nutrition, she also has written widely on nutrition for the popular press.

LOUIS W. SULLIVAN, M.D., Secretary of the Department of Health and Human Services. Prior to accepting this appointment, Dr. Sullivan served as President and Dean of The Morehouse School of Medicine. Dr. Sullivan is currently President, Association of Academic Minority Physicians, and serves on numerous advisory boards, including the Robert Wood Johnson Health Policy Fellowship Board, Institute of Medicine, National Academy of Sciences. His numerous awards include The Equitable Black Achievement Award in Education and the Distinguished Community Service Award of the Atlanta Urban League.

M.S. SWAMINATHAN, Ph.D., President, International Union for Conservation of Nature and Natural Resources (IUCN). He previously served as Director General of the International Rice Research Institute in the Philippines. Dr. Swaminathan has received numerous national and international awards, including the Albert Einstein World Science Award, the first World Food Prize, and the first Golden Heart Award given by the Philippine government. He is the recipient of thirty honorary doctorates.

R. THOMAS VYNER, Joint Managing Director, J. Sainsbury, PLC. Mr. Vyner's career in the food industry includes 23 years with Allied Suppliers including posts with Lipton and Main Board Director. Thereafter he joined J. Sainsbury PLC, where he has served as Assistant Managing Director, Buying and Marketing, and Deputy Chairman of Homebase Limited in addition to his current position.

CLAYTON YEUTTER, J.D., Ph.D., Secretary of the U.S. Department of Agriculture. Dr. Yeutter's prior government service included U.S. Trade Representative, Assistant Secretary for International Affairs and Commodity Programs, USDA, and Assistant Secretary for Marketing and Consumer Services, USDA. His private sector accomplishments include serving as President and Chief Executive Officer of the Chicago Mercantile Exchange and senior law partner with the law firm of Nelson, Harding, Yeutter and Leonard.

FRANK E. YOUNG, M.D., Ph.D., Commissioner of the Food and Drug Administration. Dr. Young was previously Chairman of the Department of Microbiology, Dean of the School of Medicine and Dentistry, and Vice President of Health Affairs at the University of Rochester. He is a member in the Institute of Medicine of the National Academy of Sciences. He also has served on a number of advisory committees of the National Institutes of Health and the American Cancer Society. He is the recipient of four honorary degrees.

Index

Academic research, 67-69
Access to food, nutrition security and, 31. *See also* Hunger
Africa, demography and, 14
Age level, demography and, 13-14. *See also* Elderly
Agricultural chemicals:
  biotechnology and, 94-95
  disadvantages of, 45
  public education and, 109
Agriculture:
  cultural development and, 75
  Europe and, 115, 117
  genetic engineering of plants and, 45-48
  globalization and, 91
  India, 32-33
  Mexico, 127-128
  United States history and, 25-30
  United States paradigms of, 19-24
  United States spending on research in, 27
AIDS, 42, 104
Alar, 93, 104
  consumer and, 120
  fear of, 6, 85, 90
Anderson, Philip W., 133, 141-147, 165
Animal breeding:
  genetic engineering and, 3, 49-53
  historical perspective on, 50
  India, 32-33
Animal health, genetic engineering and, 52-53
Anthropic principle, nutrition and, 73-74
Apples, *see* Alar
Arrow, Kenneth J., 133, 135-140, 165
Asceptic packaging, food industry and, 154
Asia:
  demography, 14
  hunger in, 32
Australia, investments in U.S. by, 155

Bacon, Francis, 108
Bak, Per, 145
Bangladesh, famine in, 14

Bateman, Durward F., 11, 19-24, 165
Beef import ban, Europe, 95, 111. *See also* International trade
Belausteguigoitia, Eneko de, 112, 127-128,165-166
Bell, Daniel, 11, 13-18, 141, 166
Berghorst, D. Theodore, 112, 123 125, 166
Biodegradability, 121-122
Bioengineering, *see* Biotechnology
Biological science, revolution in, 2-3
Biomass technology, 34
Biotechnology, 2-3
  animals, 49-53
  branding and, 152-154
  capital investment and, 123-125
  chemistry and, 61
  communication gap and, 5-7, 90, 94-95
  consumer and, 120
  developing countries and, 33, 34-35
  disenchantment with, 158-159
  DNA molecule and, 41-44
  Europe, 129
  fear of, 6
  food industry and, 159
  food safety and, 152
  gene loss and, 32
  global food system and, 151
  governments and, 156
  health and, 59-60
  hunger and, 97
  impact of, 2-3
  industry and, 39-40
  media and, 151
  overview of, 37-38
  patents and, 7-8
  plants, 45-48
  politics and, 89, 91-95
  public education and, 90
  regulation and, 7, 101-104
  research and, 67-68, 91
  safety issues and, 47-48, 53
  speed of developments in, 151-152

technology transfer and, 7-8, 35
  United States agriculture and, 27-28
Body chemistry, individual differences, 5, 56,
  64
Bovine somatotropin (BST), 28, 44, 89, 90
  attacks on, 92-93
  branding and, 153-154
  consumer and, 120
  economic dislocation and, 29, 93-94
  fear of, 6, 23, 25, 48
  regulation and, 101-102
  Soviet Union and, 150-151
  United Kingdom and, 121
Branding, food industry and, 152-154
Brazil, soils of, 92
Breast cancer, genetic engineering and, 51
Breeding, see Animal breeding
Brill, Winston J., 45-48, 166
Britain, see United Kingdom
Brown, George E., Jr., 11-12, 25-30, 166
Burma, demography and, 15
Business, research and, 67-69. See also Capital
  investment; Food industry

Canada:
  international trade and, 112
  investments in U.S. by, 155
Cancer, 104
  nutrition and, 64-65
  regulatory future and, 106
Capital investment, 112, 123-125
  biotechnology and, 159
  food industry structure and, 154-155
  global food system and, 152
Cattle, see Animal breeding
Cereals, branding and, 152-153
Ceres Conference, purposes of, 1
Chemical additives, diet and, 85
Chemical pesticides, see Agricultural chemicals
Chemicals, regulatory future and, 105
Chemistry:
  limits of, 71-72
  nutrition and, 55-56, 59-69
Chernobyl disaster, 16
China, 143, 144
  demography and, 15
  ecological resources and, 32
Cholesterol:
  debate over, 4-5
  diet and, 62
  eggs and, 153
  health and, 62-63
  labeling and, 97-98
  public education and, 99
Chromosomes, DNA molecule and, 41-44
Churchill, Winston, 118
Climate, see also Environmentalism
  genetic engineering and, 46
  global food system and, 152
Cockfield, Arthur, 114
Coloring agents, genetic engineering and, 47

Combustion, metabolism and, 72
Communication gap, see also Media; Public
  education
  agricultural technology and, 23
  biotechnology and, 5-7, 90, 91, 94-95
  marketing and, 112
  risk and, 56-57
Competition, biotechnology research and, 92
Consolidation, food industry and, 155
Consortiums, food industry and, 157-158
Consultative Group for International
  Agricultural Research (CGIAR), 159
Consumer:
  confusion among, 89, 97, 98
  food industry and, 119-120, 156
  France, 129
  global food system and, 152
  marketing and, 111-112
  nutrition and, 83-87
  United Kingdom, 121
Contaminants, regulatory challenge of, 7
Cooking oils, see Oils
Corn, genetic engineering and, 2-3
Crick, Francis, 2, 42
Culture:
  agriculture and, 75
  globalization and, 131
Cyanide, grape contamination by, 104

Dairy cattle, see Animal breeding; Milk
Daniels, George, 24
David, Paul, 138
deGaulle, Charles, 114
Delors, Jacques, 114
Demography:
  consumer and, 84
  developing nations and, 3-4
  elderly and, 79
  food industry and, 156
  forecasting and, 136
  global food system and, 151
  Mexico, 127
  perspective on, 13-18
Denman, Roy, 111, 113-118, 166
Developing countries:
  demography and, 3-4
  ecology and, 32
  genetic engineering and, 97
  genetic ownership and, 46
  politics and, 26
  technology requirements and, 32-34
  technology transfer and, 35
  vitamin deficiency and, 65
Diabetes, evolution of diet and, 77
Diet. See also Nutrition
  cholesterol and, 62
  confusion over guidelines for, 4-5
  elderly and, 80-81
  government and, 84
  health and, 62-63, 89-90
  nutrition and, 73

prehistory and, 75
regulatory future and, 106-107
Disease, *see* Health
DNA molecule. *See also* Biotechnology
genetic engineering and, 51, 53
structure of, 2, 41-44
Dunlop, John T., ix-xi, 166-167

Earthquake, 144-146
Ecology, *see* Environmentalism
Economic factors. *See also* Developing countries
biotechnology and, 94
Europe, 115, 117-118
food safety and, 92-93
future predictions, 133, 135-140, 149
Mexico, 128
Education, *see* Communication gap; Media;
Public education
Eggs, cholesterol levels in, 153
Elderly:
consumer and, 84
demography and, 13
marketing and, 129
nutrition and, 5, 56, 79-82
Ellul, Jacques, 28-29
Emigration, *see* Immigration and emigration;
Migration
Employment, developing countries,
33-34
Energy resources, future prediction and, 142
Environmentalism, 16-17. *See also* Climate
bovine somatotropin (BST) and, 102
Europe and, 130
global food system and, 152
hunger and, 32
United Kingdom and, 121-122
Enzymes, research in, 59, 60
Ethics, biotechnology and, 28
Ethiopia, famine in, 15
Europe:
beef import ban by, 95, 111
bovine somatotropin (BST) and, 102
demography, 13, 15-16
economic unification of, 111
food industry and, 129-131, 158
international trade and, 113-118
regulation in, 129-130
Evolution, dietary change and, 76-78

Famine, *see also* Hunger
demography and, 14-15
physiological protection against, 74
Farm labor, United States statistics on, 20, 25.
See also Agriculture
Feeds, genetic engineering and, 46-47
Fertilizer, genetic engineering and, 46
Feynman, Richard, 41
Fierz, Marcus, 62
Finance:
Europe, 118
global food system and, 151

Flavoring agents, genetic engineering and, 47
Food additives, regulatory future and, 107
Food industry:
biotechnology and, 159
branding and, 152-154
consortiums and, 157-158
consumer and, 119-120, 156
demography and, 156
Europe, 129-131, 158
examples of company perspectives, 156-158
France, 129-131
Mexico, 127-128
packaging and, 154
structural changes in, 154-156
United Kingdom, 121-122
Food labeling, *see* Labeling
Food regulation, *see* Regulation
Food safety:
agricultural chemicals and, 95
biotechnology and, 152
consumer and, 120
diet and, 85
health and, 92-93
regulatory future and, 105-108
United Kingdom, 121-122
Food security, nutrition security contrasted, 31
Food system, *see* Food industry; Global food
system
Forecasting, *see* Economic factors, future
predictions
Foreign markets, *see* World markets
Forrester, T. W., 141
France:
food industry in, 129-131
investments in U.S. by, 155
Future global food system, *see* Global food
system
Future predictions:
global food system and, 159-161
Mexico, 127-128
perception and, 141-147

Gabor, Dennis, 141
GATT, Europe and, 117
Gaull, Gerald E., l-9, 167
Gautier, Francis, 112, 129-131, 167
Genetic engineering, *see* Biotechnology
Germany:
demography and, 13
politics in, 6
Glaser, Donald A., 37, 39-40, 167
Global food system, 149-162
factors in, 150-152
future trends and, 159-161
past trends and, 149-150, 161
political change and, 150-151
Globalization, *see also* International trade
agricultural crisis and, 30
agriculture and, 91
branding and, 153-154
food habits and, 131

Globalization (*Continued*)
    food industry structure and, 154-155
    future trends and, 159-160
    impact of, 2
    regulation and, 7, 95, 107-108, 112
    United States agriculture and, 25
Goldberg, Ray A., ix, 149-162, 167
Government, *see* also Politics; Regulation
    biotechnology and, 156, 158-159
    diet and, 84
    food industry structure and, 155-156
    regulatory agencies of, 163
    research funding and, 66
    resource conservation and, 142
Grains, branding and, 152-153
Grain surpluses (U.S.), global food system and,
    149-150
Grapefruit, genetic engineering and, 46
Greenhouse effect, 16, 140
Green movement:
    Germany, 6
    United Kingdom, 121-122
Growth hormone, genetic engineering and, 52

Health:
    aging and, 80
    cholesterol and, 62-63
    consumer and, 119-120
    food safety and, 92-93
    future trends and, 160-161
    global food system and, 152
    nutrition and, 4-5, 56, 59-69, 89-90, 97-100
    regulation and, 105-108
    regulatory challenge and, 7
    research and, 73
Heart disease, regulatory future and, 106
Hirsch, Jules, 56, 71-74, 167-168
Human Genome Project, launching of, 2
Hunger, *see* also Famine
    genetic engineering and, 97
    international trade and, 95
    nutrition and, 65
    technological advances and, 31-36

Immigration and emigration, demography and,
    14. *See* also Migration
Immunity, genetic engineering and, 52-53
India, 3, 4
    agricultural statistics of, 32-33
    bovine somatotropin (BST) and, 151
    food availability in, 31
Individual differences:
    body chemistry, 5, 64
    nutrition and, 56, 64, 73, 78
Indonesia, 3, 4
Industry, *see* Business; Capital investment; Food
    industry
Inflation:
    biotechnology and, 94
    Mexico, 128

International Board of Plant Genetic Resources
    (IBPGR), 159, 160
International trade, *see* also Global food system;
    Globalization; Markets and marketing
    Europe and, 111, 113-118
    hunger and, 95
    Mexico, 128
    regulatory future and, 107-108
    United States agriculture and, 26-27
International Union for the Conservation of
    Nature and Natural Resources, 35
Investment, *see* Capital investment
Irradiation:
    consumer and, 120
    fears of, 23
    United Kingdom and, 121

Japan, 116
    demography and, 13, 14
    investments in U.S. by, 155
Johnson, Lyndon B., 42-43

Kissinger, Henry, 32
Kornberg, Arthur, 37, 55-56, 59-69, 168
Kretchmer, Norman, 75-78, 168

Labeling:
    Europe, 130, 131
    importance of, 90, 97
    public education and, 99
Latin America, demography and, 14
Lavoisier, Antoine Laurent, 72
Law, genetic ownership and, 46. *See* also Patents
Law of random walk, 146
Leder, Philip, 38, 49-53, 168
Limits to Growth theory, 141-142

Maquialldora area (Mexico), demography and,
    14
Markets and marketing, *see* also Consumer;
    International trade
    alar and, 93
    bovine somatotropin (BST) and, 94
    branding and, 153
    communication gap and, 112
    consumers and, 111-112, 119-120
    elderly and, 129
    food industry and, 156
    food safety and, 95
    health claims and, 98
Mechanization, migrations and, 28
Media:
    biotechnology and, 151
    consumer and, 84-85, 98
    food safety and, 120
    political attitudes of, 144
Medical education, changes in, 61
Mendel, Gregor Johann, 50
Metabolism:
    combustion and, 72

dietary change and, 76
  of milk, 5, 76-77
Methionine, 46-47
Mexico, 3-4, 112
  demography and, 14
  food industry in, 127-128
Microbial contaminants:
  regulatory challenge of, 7
  regulatory future and, 106
Microwave oven, 119
Middle East, demography and, 14
Migration, see also Immigration and emigration
  cotton picker and, 28
  global food system and, 152
Miles, Michael A , 111-112, 119-120, 168
Milk:
  bovine somatotropin (BST) and, 6, 24, 25
  genetic engineering and, 3, 50
  metabolism of, 5, 76-77
  prehistory and, 76
Miller, Sanford A., 90, 105-108, 168
Monetary policy, Europe, 118
Monnet, Jean, 113
Moore, John A., 90, 109-110, 169
Moore, Thomas, 62

National Institutes of Health (NIH), 66, 72
Nation-state, inadequacy of, 2
Natural resources:
  future prediction and, 142
  global food system and, 152
Neal, James, 77
Nonionizing radiation, regulatory future and, 107
North Africa, demography, 14
Nutrition, see also Diet
  anthropic principle in, 73-74
  chemistry and, 55-56, 59-69
  consumer and, 83-87, 119-120
  diet and, 73
  elderly and, 56, 79-82
  evolution and, 76
  future trends and, 160-161
  health and, 4-5, 56, 97-100
  historical perspective on, 71-74
  individual differences and, 73
  risk and, 56-57
Nutrition security, food security contrasted, 31

Obesity, epidemiology of, 4
Oils, genetic engineering and, 47
Organization for Economic Cooperation and
    Development (OECD), bioengineering and,
    104
Ownership, see Patents
Ozone depletion, 16

Packaging:
  Europe, 130
  food industry and, 154

Patents:
  branding and, 153
  global food system and, 152
  technology transfer and, 7-8, 35
Pesticides, see Chemical pesticides
Pest resistance, genetic engineering and, 45-46
Plants:
  genetic engineering of, 45-48
  regulation of bioengineered, 102-103
Plastics, genetic engineering and, 47
Politics, see also Government; Regulation
  biotechnology and, 89, 91-95
  developing countries and, 26
  food safety and, 120
  globalization and, 2, 150-151
  media and, 144
  utrition and, 65
  regulatory future and, 105-108
  research funding and, 66
Prediction, see Economic forecasting
Prehistory, diet and, 75-78
Probability, 136
Profits, profit margins for selected companies,
    161
Public education, see also Communication gap;
    Media
  genetic engineering and, 90
  labeling and, 99
  need for, 109-110

Radiation, regulatory future and, 107
Recombinant DNA technology, see
    Biotechnology
Recycling, public education and, 109
Regulation, 90. See also Government; Politics
  agencies responsible for, 163
  agricultural chemicals and, 95
  biotechnology and, 101-104, 158
  challenges facing, 7
  diet and, 84
  Europe, 129-130
  food safety and, 120
  future issues in, 105-108
  future trends in, 160
  globalization and, 112
  labeling and, 99
  United Kingdom and, 121
Religion, biotechnology and, 28
Research, see Agricultural research
Risk:
  communication gap and, 6, 56-57
  diet and, 85
  global food system and, 151
  regulatory future and, 107
RNA, virus and, 42
Roosevelt, Franklin D., 36
Rosenberg, Irwin H., 56, 79-82, 169
Rutan, Vernon, 27

Safety, see Food safety

Scale-free behavior, 144-146
Schuman, Robert, 113
Secrecy, research and, 67-68
Shevardnadze, Eduard, 30
Soil resources:
  Brazil, 92
  hunger and, 32
Soviet Union:
  agriculture and, 30, 91-92
  bovine somatotropin (BST) and, 150-151
  demography and, 15
Soybeans, genetic engineering and, 46-47
Starvation, physiological protection against, 74.
    See also Famine; Hunger
Stern, Judith S., 56-57, 83-87, 169
Subsidies, agricultural production and, 16
Sullivan, Louis W., 89-90, 97-100, 169
Swaminathan, M. S., 12, 31-36, 169
Sweeteners:
  branding and, 154
  genetic engineering and, 46

Tariffs, Europe and, 114, 115
Technology, see Biotechnology
Thaumatin, 46
Thomas, Louis, 66
Toxicology, regulatory future and, 107
Tropical rain forests, bioengineering and, 35

United Kingdom:
  food industry in, 121-122
  investments in U.S. by, 154-155
United Nations, 4, 14, 35, 160
United States:
  agricultural history in, 25-30
  agricultural paradigms in, 19-24
  demography and, 13
  foreign investments in, 154-155, 161
United States Department of Agriculture,
    biotechnology research and, 91
United States Environmental Protection Agency
    (EPA), bovine somatotropin (BST) and, 102
United States Food and Drug Administration
    (FDA):
  bioengineered plant regulation, 103
  bovine somatotropin (BST) and, 6, 102
  labeling and, 99
United States National Academy of Sciences, 48

Venture capital, see Capital investment
Virus, RNA and, 42
Vitamins:
  developing nations and, 65
  elderly and, 81
  research in, 59, 60
Vyner, R. Thomas, 112, 121-122, 169-170

Wall Street, see Capital investment
Waste disposal, Europe, 130
Water resources:
  global food system and, 152
  hunger and, 32
Watson, James, 2, 42
Wheat, genetic engineering and, 46
World hunger, see Famine; Hunger
World markets, see Global food system;
    Globalization; International trade; Markets
    and marketing

Yeutter, Clayton, 89, 91-95, 170
Young, Frank E., 86, 90, 101-104, 170